NATIONAL ACADEMIES | Sciences
Engineering
Medicine

NATIONAL
ACADEMIES
PRESS
Washington, DC

Chemical Terrorism

Assessment of U.S. Strategies in the Era of Great Power Competition

I0109484

Committee on Assessing and Improving
Strategies for Preventing, Countering,
and Responding to Weapons of Mass
Destruction Terrorism: Chemical Threats

Board on Chemical Sciences and
Technology

Division on Earth and Life Studies

Consensus Study Report

NATIONAL ACADEMIES PRESS 500 Fifth Street NW Washington, DC 20001

This activity was supported by Contract No. AWD-001178 between the National Academy of Sciences and the U.S. Department of Defense. Any opinions, findings, conclusions, or recommendations expressed in this publication are those of the author(s) and do not necessarily reflect the views of any organization or agency that provided support for this project.

International Standard Book Number-13: 978-0-309-70678-0
International Standard Book Number-10: 0-309-70678-5
Digital Object Identifier: https://doi.org/10.17226/27159

Library of Congress Control Number: 2024943236

This publication is available from the National Academies Press, 500 Fifth Street, NW, Keck 360, Washington, DC 20001; (800) 624-6242 or (202) 334-3313; http://www.nap.edu.

Suggested citation: National Academies of Sciences, Engineering, and Medicine. 2024. *Chemical Terrorism: Assessment of U.S. Strategies in the Era of Great Power Competition*. Washington, DC: The National Academies Press. https://doi.org/10.17226/27159.

The **National Academy of Sciences** was established in 1863 by an Act of Congress, signed by President Lincoln, as a private, nongovernmental institution to advise the nation on issues related to science and technology. Members are elected by their peers for outstanding contributions to research. Dr. Marcia McNutt is president.

The **National Academy of Engineering** was established in 1964 under the charter of the National Academy of Sciences to bring the practices of engineering to advising the nation. Members are elected by their peers for extraordinary contributions to engineering. Dr. John L. Anderson is president.

The **National Academy of Medicine** (formerly the Institute of Medicine) was established in 1970 under the charter of the National Academy of Sciences to advise the nation on medical and health issues. Members are elected by their peers for distinguished contributions to medicine and health. Dr. Victor J. Dzau is president.

The three Academies work together as the **National Academies of Sciences, Engineering, and Medicine** to provide independent, objective analysis and advice to the nation and conduct other activities to solve complex problems and inform public policy decisions. The National Academies also encourage education and research, recognize outstanding contributions to knowledge, and increase public understanding in matters of science, engineering, and medicine.

Learn more about the National Academies of Sciences, Engineering, and Medicine at **www.nationalacademies.org**.

COMMITTEE ON ASSESSING AND IMPROVING STRATEGIES FOR PREVENTING, COUNTERING, AND RESPONDING TO WEAPONS OF MASS DESTRUCTION TERRORISM: CHEMICAL THREATS

TIMOTHY J. SHEPODD (*Chair*), Sandia National Laboratories (retired)
MARGARET E. KOSAL (*Vice Chair*), Georgia Institute of Technology
GARY A. ACKERMAN, University at Albany, State University of New York
PHILIPP C. BLEEK, Middlebury Institute of International Studies at Monterey
GARY S. GROENEWOLD, Idaho National Laboratory (retired)
DAVID J. KAUFMAN, Center for Naval Analyses
KABRENA E. RODDA, Pacific Northwest National Laboratory
NEERA TEWARI-SINGH, Michigan State University
GUY VALENTE, County of El Dorado, California *(until January 2023)*
USHA WRIGHT, SHARE Africa

Staff

LINDA NHON, Study Director
ALEX TEMPLE, Program Officer
MICAH LOWENTHAL, Director, CISAC
HOPE HARE, Administrative Assistant
MARIE KIRKEGAARD, Program Officer *(until June 2022)*
MEGAN HARRIES, Program Officer *(until August 2022)*
JESSICA WOLFMAN, Research Associate *(until May 2023)*

Reviewers

This Consensus Study Report was reviewed in draft form by individuals chosen for their diverse perspectives and technical expertise. The purpose of this independent review is to provide candid and critical comments that will assist the National Academies of Sciences, Engineering, and Medicine in making each published report as sound as possible and to ensure that it meets the institutional standards for quality, objectivity, evidence, and responsiveness to the study charge. The review comments and draft manuscript remain confidential to protect the integrity of the deliberative process.

We thank the following individuals for their review of this report:

SETH CARUS, National Defense University
ROBERT CASILLAS, U.S. Army
JONATHAN FORMAN, Pacific Northwest National Laboratory
APARNA HUZURBAZAR, Los Alamos National Laboratory
FRANCES LOCKWOOD, Solar Energy Solutions LLC
SCOTT MILLER, Yale University
KATHLEEN VOGEL, Arizona State University
AUDREY WILLIAMS, Lawrence Livermore National Laboratory

Although the reviewers listed above have provided many constructive comments and suggestions, they were not asked to endorse the conclusions or recommendations of this report nor did they see the final draft of the report before its release. The review of this report was overseen by **SUSAN KOCH,** Department of State, National Security Council (retired), and **MIRIAM E. JOHN (NAE),** Sandia National Laboratories. They were responsible for making certain that an independent examination of this report was carried out in accordance with the standards of the National Academies and that all review comments were carefully considered. Responsibility for the final content of this report rests entirely with the authoring committee and the National Academies.

Contents

Summary   1

    Study Task, Scope, and Method, 2
    Complex Chemical Threat Landscape, 2
    Assessing Strategies for Identifying Chemical Threats, 3
    Strategies to Prevent and Counter Chemical WMD, 5
    Budget Recommendations, 13
    References, 15

1   Introduction   17

    1.1 Statement of Task, 17
    1.2 Chemical Environment, 22
    1.3 Report Organization, 27
    References, 27

2   Chemical Threats and U.S. Governmental and Nongovernmental
    Institutions That Play a Role (The Threat and the Who's Who)   33

    2.1 Complex Chemical Threat Landscape, 33
    2.2 Characterization of BROAD Chemical Threats, 45
    2.3 Delivery Methods of Chemical Agents, 48
    2.4 Emerging Chemical Threat Technologies, 49
    2.5 Emerging Actors, 53
    References, 56

3   Evaluation of Strategies   61

    3.1 Overview of Strategies Assessed, 61
    3.2 Methodology of Assessment, 62
    References, 64

4   Adequacy of Strategies to Identify Chemical Threats   **65**
     4.1 Analysis of Strategies to "Identify" WMDT Chemical Threats, 71
     4.2 "Identify" Strategy Efficacy, 79
     4.3 Implication of the National Strategic Shift from VEO to GPC
         from the Perspective of "Identify", 80
     4.4 Summary, 81
     References, 82

5   Adequacy of Strategies to Prevent and Counter Chemical Terrorism   **85**
     5.1 Analysis of Strategies to "Prevent or Counter" Chemical
         Terrorism Threats, 87
     5.2 Implication of National Strategic Shift From Violent Extremist
         Organizations to Great Power Competition From the Perspective of
         "Prevent/Counter", 107
     5.3 Summary, 107
     References, 108

6   Adequacy of Strategies to Respond to Chemical Terrorism   **111**
     6.1 Analysis of Strategies for "Responding" to WMDT
         Chemical Threats, 112
     6.2 Summary, 134
     References, 134

**7   Chemical Terrorism in the Era of Great Power Competition:
     Cross-Cutting Findings, Conclusions, Recommendations**   **137**
     7.1 Department of Homeland Security (DHS) Strategy, 140
     7.2 Department of Defense Strategy, 143
     7.3 Intelligence Community Strategy, 144
     7.4 Chemical Terrorism Risks, 145
     7.5 Approaches to Identify, Prevent, Counter, and Respond with Broad
         Applicability, 146
     7.6 Threat-Agnostic Approaches to Medical Countermeasures Against
         Chemical Threats, 148
     7.7 Similarities and crossover in Efforts to Counter Threats from
         Bioterrorism and Chemical Terrorism, 150
     7.8 Budget Recommendations, 151
     7.9 Summary, 155
     References, 156

Appendixes
A U.S. Government Strategies and Other Documents Considered   **159**
B Acronym/Initialism List   **167**
C Committee Biographies   **173**
D Strategy Assessment Rubric   **177**
E International Case Studies   **183**
F Threats Interdicted Case Studies   **189**
G Threats Manifested Case Studies   **195**

Preface

The strategies of the United States are exceptionally important as they influence policy, budgets, programs, and actions. Counterterrorism strategies against chemical terrorism have evolved and been supplemented since the days after 9/11/2001 as both threats and counterterrorism capabilities have evolved. Today, the United States has issued strategies that clearly prioritize great power competition (GPC) as the most important threat to world order. Terrorism has not disappeared, but it has been subordinated in prominence in U.S. national strategy. This committee was tasked with evaluating U.S. strategies against chemical terrorism during a time of overt shift in strategy to prioritize GPC over other threats. This change in national strategic priorities will result in new priorities, programs, and risks. How much national attention and resources should be given to chemical terrorism (and terrorism more broadly) as national priorities and as risk acceptance changes is a difficult question to answer. The committee took a high-level approach to this broad topic and included the advantages of various budget functions (see Table S-1).

Over the 16-month study period (January 2022–June 2023) our diverse committee met over a dozen times both in person and virtually. Regardless of the backdrop of dynamic national strategic priorities, the committee evaluated many national policy and strategy documents, some of which were issued during the course of this study group. (See Appendix Table A1 for a list of documents considered by the committee.) The committee also conducted multiple information-gathering sessions both at the National Academies and other agency locations (see Appendix Table A-2 for a list of organizations and individuals interviewed by the committee).

The committee created an evaluation rubric used to assess a subset of the National Strategies related to identifying, preventing, countering, and responding to potential chemical terrorism events (see Appendix D). A review of past, including thwarted, chemical terrorism events was conducted and any trends were analyzed. As the vast majority of toxic chemical releases come from accidents, and chemical terrorism can result from a myriad of toxic chemicals used every day, the committee considered many factors that might enable or deter terrorism including the motivations of the terrorist. A great asset against chemical terrorism is the strong first responder communities throughout the United States and established policies for escalation of chemical events.

In addition to examining strategies and the assets that can support implementation, the committee also identified obstacles to implementing strategies to prevent, counter, and respond to chemical terrorism. In particular, the failure of Congress to reauthorize the Chemical Facility Anti-Terrorism Standards (CFATS) program (6 CFR Part 27) legislation in July 2023 was noted by the committee at the time of expiration. CFATS is a coordinated federal program focusing on enhancing security measures at more than 40,000 domestic chemical facilities that help protect them from potential acts of terrorism, including insider threats and cyberattacks. CFATS reauthorization is supported by the chemical industry, the American Chemical Society, and this committee to ensure that chemical facilities operators are taking steps to reduce and mitigate the potential for terrorist exploitation of this vital critical infrastructure.

Finally, we want to give our sincerest thanks to the members of the committee, the many briefers who shared their reality of how strategy played out in their organizations, and the numerous talented professionals at the National Academies, including their IT support staff.

> Tim Shepodd, *Chair*
> Margaret E. Kosal, *Vice Chair*
> Committee on Assessing and Improving
> Strategies for Preventing, Countering,
> and Responding to Weapons of Mass
> Destruction Terrorism: Chemical Threats

Summary

Chemical threat agents are highly hazardous or toxic chemicals that can be acquired or used as weapons by states or nonstate actors. The widespread availability of starting materials and instructional materials for producing chemical weapons of mass destruction (WMD) have reduced barriers to entry for the nefarious use of chemicals. Domestic and foreign violent extremist organizations (VEOs), or terrorist groups, have caused a greater amount of harm with chemical agents than with biological or radiological weapons.

The United States' capacity and capability to identify, prevent, counter, and respond adequately to chemical threats is established by the strategies, policies, and laws enacted across multiple levels of government. Many U.S. counter-WMD terrorism policies, strategies, and programs were enacted in the wake of the September 11, 2001, terrorist attacks on the United States. Shortly after, the subsequent mailing of envelopes containing spores of B. anthracis also propelled a number of WMD-related counterterrorism and nonproliferation programs across different agencies. In addition, steady progress has been made over that same time in eliminating declared chemical weapons stockpiles under the Chemical Weapons Convention (CWC), which went into effect in 1997 and has more than 190 states/parties.

While the number of chemical terrorism incidents has risen and fallen over time, there is no empirical or analytical indication that the threat is disappearing, especially with several incidents within the past three decades of terrorists using or pursuing nerve agents or chemical agents. Factors that could potentially increase this threat include the large and growing number of chemicals that could be used in chemical weapons, perceived changes in the tactical and/or strategic benefits of using them compared to other types of weapons, emerging technologies, and a rise in foreign or domestic terrorism. It is in this context that the 2021 National Defense Authorization Act (NDAA) directed the Secretary of Defense to request that the National Academies of Sciences,

Engineering, and Medicine (NASEM) conduct an independent review of the adequacy of U.S. strategies to prevent, counter, and respond to chemical terrorism.

STUDY TASK, SCOPE, AND METHOD

Given the breadth of the study's task, the committee took a high-level view and focused on identifying the most important technical, policy, and resource gaps with respect to strategies for identifying, preventing, countering, and responding to chemical threats, and budgeting to address these needs. While acknowledging the potential rise of terrorist threats from state actors over the past decade, the report focuses on chemical threats originating from nonstate actors with or without state involvement (e.g., knowledge or the capability to share and other forms of support to enable chemical terrorism). The committee decided to combine preventing and countering in their strategy assessment. Considerations of long-term health and environmental effects were beyond the scope of the charge. To carry out its work, the committee systematically evaluated key strategic documents listed in chapter 3, that range from national-level to agency-level strategies.

In the United States, there has not been a chemical terrorist event that has had consequences approaching those observed outside of the country. Generally, U.S. response organizations have been effective in identifying and thwarting chemical threats, although, there have been a few notable cases where law enforcement did not identify a threat before an attack was executed. Additionally, the 2018 Skripal poisonings in the United Kingdom illustrate a new turning point in actors, intent, and methods in the chemical threat: from that of terrorist-initiated to use by a combination of a state actor, targeted assassination, and nontraditional agent.

COMPLEX CHEMICAL THREAT LANDSCAPE

Incidents of chemical terrorism and attempted terrorism have involved one hundred different perpetrators motivated by different ideologies (see Figure S-1). For the period between 1990–2017, the geographic distribution of countries where chemical terrorism

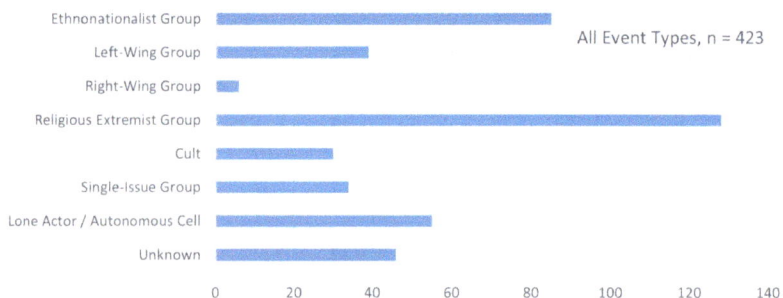

FIGURE S-1 Interest in or pursuit of chemical weapons by perpetrator.
SOURCE: Binder and Ackerman, 2020.

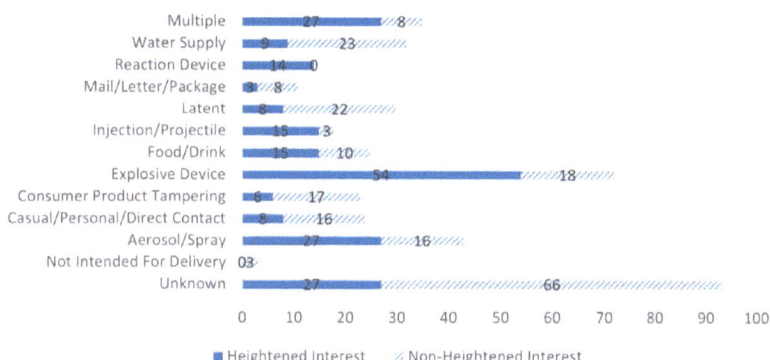

FIGURE S-2 Chemical terrorism incidents by intended delivery method.
SOURCE: Binder and Ackerman, 2020.

incidents have occurred is extensive, with 47 countries involved over this period and actual uses of chemical agents as weapons occurring in 34 of these cases. They have also utilized an array of chemical agents encompassing many commonly available (often referred to as "low-end") agents and several toxic industrial chemicals (TICs), as well as toxic industrial materials (TIMs). They have also used some chemical agents that have traditionally been developed in the military context (e.g., the nerve agent, sarin) and have included a variety of delivery methods, such as explosive devices, aerosol, and other methods (see Figure S-2).

ASSESSING STRATEGIES FOR IDENTIFYING CHEMICAL THREATS

The total number of chemicals that constitute or could constitute WMD terrorism threats is vast and continually expanding. Over two hundred million chemicals have been synthesized or isolated, and another is identified every 3–4 seconds (CAS, n.d.). Technological advances such as cheminformatics, artificial intelligence, machine learning, additive manufacturing, nanotechnology, and microscale chemical reactors further facilitate the discovery of new and novel chemical threat agents available for potential beneficial or nefarious use. Thus, it is impossible to identify and prevent or counter every threat.

Federal agencies that spoke with the committee acknowledged that, overall, terrorists seeking to perpetrate chemical attacks tend to opportunistically misuse traditional classes of chemicals, primarily TICs and TIMs. The majority of publicly reported domestic plots did not come to fruition between the 1970s and mid-2010s for several reasons. However, the occurrence or nonoccurrence of terror attacks involving chemicals is not a direct indication that the United States, in particular, the intelligence community (IC), was or was not successful in identifying a threat.

RECOMMENDATION 2-3: The intelligence community (IC) should continue to monitor interest in emerging technologies and delivery systems, such as drug

delivery systems, and trends by terrorist groups to innovate and improvise using chemical agents. This may look significantly different than the applications of advanced materials chemistry by great power states.

To assess the United States' capability to identify chemical threats, the committee reviewed recent strategies. The committee considered that a successful strategy to identify chemical terrorism threats is one that focuses on robust information-sharing regarding the following:

1. Chemicals that may be used in an attack—both known chemical weapon agents and lesser-known emerging agents;
2. Threat actors who may use or pursue chemicals for use in weapons of mass destruction terrorism (WMDT) attacks; and
3. Entities that may support or sponsor chemical attacks or terrorism.

This report concludes that the Federal Bureau of Investigation (FBI), partner law enforcement, and IC have been effective in identifying and interdicting the majority of domestic terrorist attacks involving chemical materials, which have typically employed conventional TICS rather than traditional chemical warfare agents. While the FBI has been effective, approaches to identifying chemical threats could be strengthened by using a multi-lens approach from several different agencies that emphasizes augmented communication and coordination between local and state enforcement and the IC. In addition, this area would greatly benefit from increased coordination between the IC and technical experts—particularly those with specific knowledge of terrorist motivation and psychology. Finally, it is unclear if the tactical readiness to implement the reviewed strategies is occurring at the necessary pace to respond to an act of chemical terrorism.

RECOMMENDATION 4-3 (Abbreviated): Existing IC programs should actively seek and incorporate new approaches to identify existing chemical threats (traditional and improvised) and potential emerging threats by terrorist groups. In developing new approaches, program managers should develop strategies that look beyond the traditional terrorism suspects and that augment and leverage skill sets of the U.S. Government (USG) agencies. The threat assessments should be updated reflecting the current times and demographics.

RECOMMENDATION 4-4: The National Counter Terrorism Center (NCTC), Department of Defense (DoD), and Department of Homeland Security (DHS) should review current identification approaches to determine whether shifts in emphasis are required as a result of expanded and augmented VEOs and terrorist capability resulting from the potential migration of chemical agents, other materials, technology, and expertise from state actors to VEOs.

RECOMMENDATION 4-5: The USG should ensure that the identification of chemical terrorism threats is explicitly included in ongoing and future strategies. Chemical terrorism threats should be considered distinct from nuclear nonproliferation, identification of state-based offensive chemical programs, and traditional (non-nuclear-biological-chemical) terrorism.

STRATEGIES TO PREVENT AND COUNTER CHEMICAL WMD

"Prevent/counter" strategies focus on plans to prevent and counter specific adversaries from committing acts of chemical terrorism. The committee surveyed the strategy documents listed in chapter 5, all of which contained useful information related to aspects of preventing and countering chemical terrorism. A successful strategy to prevent or counter chemical terrorism focuses on the following elements:

- Incorporates developments in the "Identify" area into practice for "Prevent and Counter."
- Dissuades terrorists through deterrence by denial, deterrence by punishment, or through normative means.
- Impedes acquisition of raw materials, production technology, delivery technology, or information for production or delivery. Strategy also demonstrates having mechanisms (e.g., insider threat programs, strategic trade controls, international efforts, collaboration with other counterterrorism programs) to ensure those items are not acquired.
- Interdicts active plots through military, law enforcement, or intelligence capabilities.
- Ensures collaboration at various levels—international, federal, state, local, tribal, and territorial.
- Addresses new chemical terrorism threats: new chemical agents, new production or delivery methods and technologies, new actors, forming collaboration with non-terrorist focused agencies, like the Drug Enforcement Agency (DEA).

The report concludes that most of the prevent/counter strategy documents espoused a coherent action plan or set of strategy elements comprising a combination of a well-defined goal with a corresponding definition of success, as well as at least one policy, plan, and/or resource allocation designed to meet the goal.

Deterrence

Deterrence is an influence strategy that tries to dissuade the adversary from undertaking some action through the use of negative incentives. Deterrence most commonly refers to the use of conditional threats, where the costs threatened are intended to outweigh the benefits from the action being considered. The committee found multiple existing policies and programs that contribute to a strategy of deterrence by denial, which involves denying attainment of benefits so that the actor is dissuaded from attempting the action in the first place. These include facility security improvements under the Chemical Facility Anti-Terrorism Standards (CFATS)[1] program and a variety of response capabilities that would mitigate the harm caused by a chemical attack.

Upon reviewing existing strategy documents, the committee found references to deterrence by punishment in a nonspecific context. For example, the 2002 National Strategy to Combat Weapons of Mass Destruction (p.3) reiterates the declaratory policy that

> the United States will continue to make clear that it reserves the right to respond with overwhelming force—including through resort to all of our options—to the use of WMD against the United States, our forces abroad, and friends and allies... posing the prospect of an overwhelming response to any use of such weapons.(Executive Office of the President, 2002)

The overall document explicitly cites terrorists as a source of potential risk in context of acquisition and use of WMD, but they are not explicitly called out in context of deterrence.

Strategies addressing nonstate actors appear to be focused predominantly on other forms of deterrence, which could involve threatening to punish potential states, nonstate institutions, and even individuals who might support terrorists acquiring WMD (including chemical weapons). The committee found no explicit declaratory statement of direct deterrence by punishment directed toward terrorists who used chemical weapons, in contrast to both the nuclear and biological domains.

There are substantial advantages to an explicit communication of the direct deterrence proposition (e.g., that the United States will take certain measures if terrorists utilize chemical weapons that would not otherwise be taken). Careful consideration should be given to incorporating direct deterrence of chemical terrorism into existing counter-WMD terrorism strategies.

RECOMMENDATION 5-1: The National Security Council should give careful consideration to incorporating direct deterrence of chemical terrorism into existing Chemical WMDT strategies.

Reducing Material Availability and Chemical Substitution

Chemicals are on a spectrum from extremely accessible (e.g., commercially available household chemicals), relatively accessible (e.g., many so-called TICs present in chemical plants and manufacturing facilities), to extremely inaccessible (e.g. organophosphate nerve agents and many of their key precursor chemicals). In theory, any chemical can be produced from readily available precursor chemicals. However, in practice, the technical barriers to producing certain chemicals are high. In some cases, for example nerve agent synthesis, the technical barriers are extremely high. Regu-

---

[1] At the time of writing this report, the statutory authority for the CFATS program (6 CFR Part 27) expired and has yet to be reauthorized.

latory efforts to reduce material availability include the Environmental Protection Agency's (EPA) Management Program Regulation, which aims to "reduce the likelihood of accidental releases at chemical facilities, and to improve emergency response activities when those releases occur." (Final Amendments to the Risk Management Program (RMP) Rule, 2018). DHS's Cybersecurity and Infrastructure Security Agency's (CISA) established CFATS, which once played an important role in regulating material availability.

Another key avenue by which the risk of chemical terrorism can be reduced is to replace existing processes and materials with less toxic alternatives, often referred to as inherently safer technology. This obviously reduces the potential consequences of a chemical terrorist attack by making toxic materials less prevalent or by eliminating their use entirely. Overall, theft and use of the materials in commerce will become more difficult and less attractive. The likelihood of a chemical facility becoming a target for sabotage will also decrease. Occupational and environmental safety concerns have long driven industry to seek substitution as a strategy to mitigate hazards, and both Occupational Safety and Health Administration (OSHA) and EPA have for decades encouraged and recognized innovative approaches for substitution. However, despite ongoing industry practice and some initiatives that previously operated under DHS's CFATS program, the strategy documents reviewed by the committee do not cite chemical substitution as a key part of an overall chemical security strategy.

RECOMMENDATION 5-3: Substitution of safer alternative chemicals for hazardous chemicals in industrial and academic settings should be included as part of the overall strategy to impede acquisition of raw materials for chemical terrorism. The planning and development of these strategies should be spearheaded by DHS's Chemical Information Sharing and Analysis Center under a reauthorized CFATS program and should continue to be conducted in conjunction with regulatory agencies, specifically, the EPA, OSHA, and representatives from industry and academic research environments.

Addressing Insider Threats

In certain sectors—often related to the materials consumed or produced therein—the threat lies not only in the theft of information and the disruption of an organization's functions, but also in the possibility that sabotage by insiders could have extremely detrimental consequences for broader public health and safety. The accidental release of more than 40 tons of highly toxic methyl isocyanate from the Union Carbide insecticide plant in Bhopal, India in 1984 is an example of the scale of harm that could result from an accident occurring at a chemical facility (Broughton, 2005, Eckerman, 2005).

Despite this significance, strategic documents surveyed did not explicitly mention insider threat in the chemical terrorism context. While CFATS included some practical efforts to counter insider threats within the chemical industry, the scope of these efforts appears to be limited. **The committee did not find evidence of a similar program at the level of CFATS, either directed towards government facilities or**

within academic research institutions. Nonetheless, government and academic institutions are subject to research security requirements, including reliability programs or controlled access to chemicals via security clearances, which do have an active insider threat identification scope. The severe consequence of an insider at a chemical facility conducting or assisting an attack warrants explicit inclusion in existing strategies and comprehensive policies as a way to counter insider threats at any facility containing significant quantities of toxic chemicals.

> **RECOMMENDATION 5-4: Counter-insider threat activities should be incorporated explicitly into broader counter WMD strategy. The DHS should develop a strategy to ameliorate insider threats explicitly for the chemical domain.**

Other Prevent and Counter Activities

Some activities the USG is undertaking are not mentioned in the strategy documents reviewed, including: military capabilities to provide early warning of chemical terrorism plots; law enforcement capabilities to counter chemical threats tactically; integration with broader counterterrorism and counter-smuggling efforts; and involvement with other multilateral activities beyond the Organization for the Prohibition of Chemical Weapons (OPCW). The absence of such activities from the strategies could impact policy implementation, such as budgeting, program prioritization, and other consequences. Including these activities in existing strategies would bolster their effectiveness.

> **RECOMMENDATION 5-5: Agencies should work to reconcile operational practice with policy by supplementing extant strategies to include current omitted effective activities and programs for countering chemical terrorism. This would ensure that effective practices are maintained, properly resourced, and reflected in comprehensive strategies.**

Adequacy of Strategies to Respond to Chemical Terrorism

The vast majority of chemical incidents in the United States are not from terrorism, but are instead chemical releases from accidents, transportation incidents, or the results of natural phenomena, which over the period of 2012–2022 caused nearly one hundred recorded fatalities and almost two thousand injuries. When accidents occur, first responders have tools, training, and interagency agreements generally adequate for protecting the U.S. population, themselves, and the environment. The EPA is the primary agency coordinating response to such incidents, with support from several other agencies.

In this study, response to a chemical terrorism event is defined as the ability to minimize effects, sustain operations, and support follow-on actions. To assess the nation's ability to respond to chemical terrorism, the committee reviewed the documents shown in chapter 6 and assessed response strategies for their ability to address the following questions:

1. Does the U.S. strategy adequately enable response capabilities (e.g., operations coordination, information-sharing, medical support, and others) that minimize potential impact to life, property, and the environment?
2. Is the strategy for responding to chemical terrorism, and the resources devoted to implement the strategy, aligned with the priorities of the United States (e.g., protecting the homeland, ensuring economic security, maintaining military strength) and aligned with the nation's risk posture?
3. Does the strategy anticipate emerging threats by suggesting the scientific research and interagency relationships necessary to respond to future threats?

The committee concluded that the current set of U.S. strategies, operational plans, and other resources has helped establish a network of capable first responder communities prepared for various chemical incidents regardless of their cause. However, improvements are needed in the following areas: need for first responder input, access to intelligence, information flow, and interagency coordination.

Need for First Responder Input

A major component for creating a robust strategy is to ensure critical information is collected and included from the first responder community. CISA released the Aviation Safety Communique (SAFECOM) Nationwide Survey to collect data from organizations that use technology for public safety, including emergency communication centers, emergency management, law enforcement, emergency medical personnel, and fire and rescue professionals. These types of input from relevant stakeholders in the response community will also eventually shape the direction of risk assessments.

Access to Intelligence

One concern raised in agency briefings is that information that would be most beneficial to first responders sometimes cannot be transmitted due to classification status of the information. At the recommendation of the 9/11 Commission, NCTC created a mobile app ACTknowledge, that shares unclassified counterterrorism reports, analysis, training resources, and alerts to users, however, as of January 2023 ACTknowledge was discontinued. Under National Institutes of Health (NIH), the National Library of Medicine hosts the mobile app and web-based platform Wireless Information System for Emergency Responders (WISER), which is designed to provide first responders with quick access to critical information during hazardous material incidents and other emergencies, but WISER was also discontinued in February 2023. Other emergency management tools are still operational, like Computer-Aided Management of Emergency Operations (CAMEO) and Chemical Hazards Emergency Medical Management (CHEMM) (EPA, n.d.; U.S. Department of Health & Human Services, n.d.). While the FBI is actively engaged in fostering communication with state and local first responders, including the National Guard, and industry, it is not clear that the outreach is comprehensive or systematic. The risk is that an event could occur in an area where first responders would not be aware of, or in communication with FBI personnel or capability.

Top-Down and Bottom-Up Information Flow

All of the briefings received by the committee from various agencies demonstrated a clear understanding of looking upstream to current authorities, strategies, policies, and laws governing internal agency responsibilities. Less clear to the committee is how the requirements systematically flow downstream from higher-level policy to subsidiary organizations and finally to first responders. For example, roles and responsibilities of EPA and DHS/Federal Emergency Management Administration (FEMA) officials as well as their chain of communication could lead to confusion at the local level; the result, a potentially slower response to a chemical incident or attack.

The main framework employed by DHS FEMA to coordinate and respond to emergency, natural disasters, or terrorist events is the National Incident Management System (NIMS), within which is the National Response Framework (NRF) with two documents related specifically to responding to chemical incidents: ESF#10 and the Oil/Chemical Incident Annex. The committee found that the NRF adequately addressed chemical terrorism, but that translating U.S. strategies and frameworks into operational practice for chemical terrorism response remains a challenge.

Enhancing Interagency Coordination

Coordination among the different organizations can be improved to ensure first responders receive the needed information.

With respect to addressing chemical attacks specifically, FEMA's WMD Strategic Group Consequence Management Coordination Unit coordinates with other parts of FEMA through its Chemical Biological Radiation and Nuclear (CBRN) Office. The FBI has designated WMD coordinators in its fifty-six field offices with the idea that building strong working relationships in place encourages a smoother response to a chemical incident. They routinely host WMD workshops to train first responders in recognizing the use of WMD during the initial stages of an incident.

> **RECOMMENDATION 6-6: Considering the complexity of the chemical threat space and USG coordination required for an effective response to a chemical event, the committee recommends continuing a robust program of interagency exercises and trainings that practice communication and resource sharing.**

Priorities in a Shifting Threat Landscape

This report evaluating U.S. strategies to address chemical terrorism comes at a time when the nation's highest-level strategies have shifted from focusing primarily on VEOs to focusing more on the Great Power Competition (GPC). In the words of President Biden,

The most pressing strategic challenge facing our vision is from powers that layer authoritarian governance with a revisionist foreign policy [. . .] a challenge to international peace and stability. (NSS, Pg. 8)

This change indicates a shift in relative perceived threat and consequent prioritization, which will impact efforts against chemical terrorism. Changes in strategy will lead to changes in funding priorities.

While changes in funding priorities and operational adjustments are anticipated, the specific mechanism, magnitude, and timing are currently less understood. If federal agencies prioritize broadly applicable approaches to all areas of the Chemical WMDT enterprise, it will maximize the USG capacity for appropriate response.

RECOMMENDATION 7-1: The shift in the global threat landscape has led to a corresponding shift in countering WMD to a focus on GPC, but care should be taken to ensure that existing capabilities focused on countering terrorism are maintained. Recommendations based on revised risk assessments that are aligned with new national-level priorities should be developed.

DHS Security Strategies

How the strategic shift from VEOs to GPC will impact DHS's strategic posture, programs, human resources, and missions is yet to be fully understood. The department has not yet published a strategy that both acknowledges the shift to the GPC and addresses chemical terrorism. Their 2020–2024 Strategic Plan does not specifically acknowledge GPC as a top national threat; in contrast, the 2021 China Strategic Action Plan (SAP) acknowledges the shift to the GPC but does not discuss chemical terrorism.

RECOMMENDATION 7-2 (Abbreviated): DHS should develop strategies, including an updated chemical defense strategy that consider the implications of the strategic shift to great power competition, including potential resourcing shifts, on reducing the risk of chemical threats and chemical terrorism.

DoD Strategies

The shift to GPC also impacts the DoD, though differently than the domestically focused DHS. DoD's intersection with chemical terrorism is part of a broader concern about terrorism threats against the United States' assets—and those of our allies—overseas and about terrorist assets that might mature into a threat against the homeland. In the National Defense Strategy (NDS), DoD embraces the shift to prioritizing GPC and will likely lead to a reallocation of resources supporting the new prioritization.

RECOMMENDATION 7-3: DoD should monitor risks associated with the shift in strategic focus and adapt if evidence of terrorist activities ramps back up.

Despite the changes in DoD and national strategy, it is not apparent how operations have adjusted to the new strategies nor is it clear how the IC will address information gaps.

> **RECOMMENDATION 7-4 (Abbreviated): The IC and its offices throughout the departments with significant chemical terrorism roles and responsibilities (DoD, DHS, Department of Justice (DOJ)) should take steps to ensure that counter chemical weapons programs, whether state-based or by nonstate actors, are not technologically deterministic.**

Chemical Terrorism Risks

If GPC intensifies, there are potential implications for chemical terrorism threats beyond a possible reduction of resources available to address the threats. The decisions that states make may wittingly or unwittingly lead to a dramatic increase in the sophistication of chemical terrorism, in terms of both the agent employed and/or the means by which it is delivered (see data from Figure S-2). States might also choose to engage in offensive chemical weapons activities (as some, notably Russia, are suspected to be doing today) and technology, materials, expertise, and/or chemical agents might be illicitly transferred or diverted to nonstate actors. In addition, there is the potential for what might be categorized as "state terrorism"—as some have alleged both Russia and North Korea have done with targeted attacks in recent years. In the opinion of the committee, these factors could lead to reduced resources for countering weapons of mass destruction terrorism (CWMDT) broadly, although the mechanisms, magnitude, and timing are currently poorly understood.

> **RECOMMENDATION 7-5: DoD should conduct risk and threat assessments to understand how best to direct resources to address risks of chemical terrorism events in an era of GPC-focused strategies.**

At the time of writing this report, the committee learned that CFATS's statutory authorization was allowed to expire. Therefore, the CISA cannot enforce compliance with the CFATS regulations at this time.

> **RECOMMENDATION 7-6: Congress should immediately reauthorize the CFATS program and consider long-term reauthorization.**

Threat-Agnostic Approaches to Medical Countermeasures

If resources for counterterrorism decrease due to the shift towards GPC, then a burden will be placed on existing programs to use their resources more efficiently in countering chemical threats. Despite the potential loss of focus on chemical terrorism, the growing trend toward more broadly extensible strategies being implemented by many agencies may help reduce risk. DoD's Chemical and Biological Defense Program (CBDP) and Biomedical Advanced R&D Authority (BARDA) have prioritized five diagnostic toxidromes for chemical exposures (neurologic, pulmonary, respiratory, metabolic, vesicating), which bypasses the need to identify the specific agent. This has positioned BARDA to more readily develop and deploy effective chemical medical countermeasures across multiple sectors to "treat the injury, not the agent." (Chemical Medical Countermeasure Overview, 2024)

RECOMMENDATION 7-7: Federal agencies should prioritize broadly applicable approaches beyond the specific mission sets represented by the U.S. Army Combat Capabilities Development Command Chemical Biological Center (DEVCOM CBC), BARDA, and CISA, to all areas of the CWMDT enterprise to maximize the United States' government capacity for appropriate response on time scales of relevance.

BUDGET RECOMMENDATIONS

The committee heard from several briefers that budgets are inadequate to address the breadth of possible chemical threats, even for agencies for which WMD is the highest priority. The material reviewed by the committee showed insufficient detail to allow a robust assessment of budgets likely to be required to implement strategies effectively, particularly for offices whose missions cover both chemical and biological threats. Revised risk assessments are needed to reprioritize risks guided by new strategies, so that strategy-aligned budgets can be created. To ensure a balance between different efforts as a result of risk assessments, as alluded to in section 1.1.2, a distinction between countering chemical and countering biological efforts is needed.

RECOMMENDATION 7-8: WMD budgets should be aligned with evolving strategic priorities.

RECOMMENDATION 7-9: Chemical WMDT budgets should incentivize activities that transition promising research to operations.

The committee recommends that chemical terrorism risk assessments (e.g., full risks, threats only, national-level, state-level, and others) be performed in the context of the latest strategies to align budget priorities with strategic priorities, and most clearly understand where and why the United States is accepting risk. Table S-1 shows the budget functions and resources the committee believes should be considered under budgetary constraints that may result from the national strategic shift to GPC. These factors include risk priorities that are expressed in budget requests.

TABLE S-1 Recommended Budget Priorities Based on National Strategic Shift to GPC

Budget Function or Resource	Benefit of Retention
Fund comprehensive risk assessments based on the priorities set forth in recent national security strategies.	Allows forward-thinking strategic planning and preparedness. Enables agility to focus on new priorities when national strategy evolves. Identifies alignment between funding emphasis and strategy. Identifies where risk is being accepted when alternate, more strategy-aligned, investments are made.
Maintain the intelligence community's capabilities and expertise specific to terrorist groups (VEOs & Racially, Ethnically, and Motivated Violent Extremists (REMVEs) and to understanding their motivations.	Ensures subject matter expertise in the terrorism threat space is retained. Allows for rapid identification of and adaptation to emerging threats.
Support basic scientific and social science research specifically related to countering chemical terrorism, e.g., understanding social behavior related to emerging threats.	Retains a strong talent base to address future, perhaps unanticipated, chemical threats/substances and the motivations to use them. Threats change and without natural and social scientific research, it will be difficult to adapt to changes, or in some cases, even understand that and/or why they have occurred.
Strengthen insider threat programs related to physical, cyberphysical, and cybersecurity across the chemical industry.	Secures physical facilities from being subverted to cause toxic releases or the theft of precursor chemicals. Protects vulnerable information systems from being used in espionage and for chemical attacks.
Support training and exercises to advance international chemical security priorities through continued initiatives with, for example, the Organization for the Prohibition of Chemical Weapons (OPCW), Proliferation Security Initiative (PSI) partners, North Atlantic Treaty Organization (NATO) allies, nongovernmental organizations, and other international stakeholders.	Increases capacity and tactical readiness internationally, thereby decreasing global threat and decreasing reliance on U.S. assets to respond.
Fund initiatives that work with international partners to enhance chemical security, identify, prevent/counter, and respond to chemical threats worldwide.	Strengthens alliance and builds stronger communication networks among relevant international agencies.
Continue emphasizing programs employing threat-agnostic approaches to identify and respond to chemical attacks.	Enables more economical, efficient, and effective responses, especially in times when chemical terrorism, or other national security concerns, may be deemphasized.
Encourage more flexible capability portfolio management models and processes that reduce bureaucratic constraints to accelerate adoption of emerging technologies. Utilize innovation like the cross-functional team program management approaches model.	Enables the flexibility to most promptly address evolving threats and to more effectively facilitate innovation adoption and integration. (Esper and Lee James, 2023)

REFERENCES

Binder, M., and G. Ackerman. 2020. "Profiles of Incidents Involving CBRN and Non-State Actors (POICN) Database. National Consortium for the Study of Terrorism and Responses to Terrorism: College Park, MD.

Broughton, Edward. 2005. "The Bhopal Disaster and Its Aftermath: A Review." *Environmental Health* 4: 6.

CAS. (n.d.). CAS registry. https://www.cas.org/cas-data/cas-registry.

Eckerman, Ingrid. 2005. *The Bhopal Saga—Causes and Consequences of the World's Largest Industrial Disaster*. India: Universities Press.

Environmental Protection Agency (EPA). (n.d.). CAMEO Chemicals. https://www.epa.gov/cameo.

Executive Office of the President. 2002. National Strategy to Combat Weapons of Mass Destruction.

U.S. Department of Health & Human Services. (n.d.). 2002. CHEMM. https://chemm.hhs.gov.

1

Introduction

Chemical terrorism is a threat because toxic chemicals and their precursors are sought and have been used by domestic and foreign violent extremist organizations (VEOs) referred to as terrorist groups. Many U.S. counter-weapons of mass destruction (CWMD or WMD) terrorism policies and strategies were enacted and received significantly more attention following the September 11, 2001, terrorist attacks on the United States and the subsequent mailing of envelopes containing spores of *B. anthracis,* the causative agent of anthrax. The use of chemical weapons by the Assad regime in Syria and violent extremist organizations as part of the Syrian civil war also brought renewed attention to the risk of chemical threats (the White House, 2012). The United States' capacity and capability to identify, prevent, counter, and respond adequately to chemical threats is established by the strategies, policies, and laws enacted across multiple levels of government.

1.1 STATEMENT OF TASK

Recognizing the need to understand current U.S. strategies to adequately address chemical terrorism, section 1299I of the 2021 National Defense Authorization Act (NDAA) directed the Secretary of Defense (who delegated to the Office of the Undersecretary of Defense for Policy OUSD(P)) to sponsor the National Academies of Sciences, Engineering, and Medicine (NASEM) to conduct an independent review of strategies to prevent, counter, and respond to chemical terrorism. Box 1-1 provides the statement of task for which the committee was charged with addressing those points.

BOX 1-1
Study Statement of Task (SOT)

The National Academies of Sciences, Engineering, and Medicine (NASEM) will appoint an ad hoc topical committee to address specific issues related to chemical terrorism threats. This committee will address the adequacy of strategies to prevent, counter, and respond to chemical terrorism, and identify technical, policy, and resource gaps with respect to:

1. identifying national and international chemical risks, and critical emerging threats;
2. preventing state-sponsored and non-state actors from acquiring or misusing the technologies, materials, and critical expertise needed to carry out chemical attacks, including dual-use technologies, materials, and expertise;
3. countering efforts by state-sponsored and non-state actors to carry out such attacks;
4. responding to chemical terrorism incidents to attribute their origin and help manage their consequences;
5. budgets likely to be required to implement effectively such strategies; and
6. other important matters that are directly relevant to such strategies.

NASEM will produce a consensus report and may produce additional products (such as proceedings of workshops) by mutual agreement with the sponsor. The consensus report will be unclassified with a classified annex.

1.1.1 Study Scope

Given the breadth of the study's statement of task, the committee has taken a high-level view of this tasking and focused on identifying the most important technical, policy, and resource gaps with respect to strategies for identifying, preventing, countering, responding to, and budgeting for chemical threats and attacks against U.S. interests. The committee decided to combine preventing and countering terrorism in their strategy assessment. Identifying emerging threats was limited to those that enable capabilities to respond to chemical terrorist attacks and their immediate effects. Table 1-1 provides definitions for several key terms used throughout the report.

Considerations of long-term health and environmental effects were beyond the scope of the charge. The committee considered both high-level approaches as well as publicly available strategy documents developed by the U.S. government (USG) and limited the

TABLE 1-1 Key Definitions

Term	Definition
Chemical Weapon	A toxic chemical and its precursors or a munition, device, or equipment specifically designed to cause death or other harm through toxic properties of those toxic chemicals. (Condensed from 18 USC Ch. 11B Chemical Weapons 229F).[a]
Chemical Terrorism	The unlawful use of chemical hazards/agents/weapons or threat of use of chemical hazards/agents/weapons against persons, property, environmental, or economic targets, to induce fear or to intimidate, coerce, or affect a government, the civilian population, or any segment thereof, in furtherance of political, social, ideological, or religious objectives (DHS, 2017). (Adapted from DHS risk lexicon).
Domestic Terrorism	Involves an act that: 1) is dangerous to human life or potentially destructive of critical infrastructure or key resources, and is a violation of the criminal laws of the United States or of any State or other subdivision of the United States; and 2) Appears to be intended to: • intimidate or coerce a civilian population; • influence the policy of a government by intimidation or coercion; or • affect the conduct of a government by mass destruction, assassination, or kidnapping. (DHS lexicon).[a]
Domestic Violent Extremist	A domestic violent extremist (DVE) is defined as an individual based and operating primarily within the United States or its territories without direction or inspiration from a foreign terrorist group or other foreign power who seeks to further political or social goals, wholly or in part, through unlawful acts of force or violence dangerous to human life.[c]
Weapon of Mass Destruction (WMD)	Chemical, biological, radiological, or nuclear weapons capable of a high order of destruction or causing mass casualties (Department of Defense [DoD] dictionary) (DoD, 2021).
Weapon of Mass Effect	Chemical, biological, radiological, or nuclear weapons capable of inflicting significant destructive, psychological, and/or economic damage to the United States (Adapted from Weapons of Mass Effect Task Force, 2006).
Emerging Threats	Threats with the potential to materialize in the next five to ten years.
Strategies	Statements of goals to fulfill assigned missions based on existing and expected resources.
Prevention	Activities and operations to dissuade states or non-states from pursuing the development of acquisition of WMD (JP 3-40).
Countering	Activities and operations to interdict or stop a chemical terrorism plot or attack that is an immediate threat or underway/being executed.
Response	Immediate actions to save lives, protect property and the environment, and meet basic human needs; include the execution of emergency plans and actions to support short-term recovery (DHS risk lexicon).

[a] See https://uscode.house.gov/view.xhtml?path=/prelim@title18/part1/chapter11B&edition=prelim#:~:text=(1)%20Chemical%20weapon%20.consistent%20with%20such%20a%20purpose.

[b] Definition of "domestic terrorism" from the Homeland Security Act definition of "terrorism," 6 USC § 101(18), which is similar, but not identical, to the 18 USC § 2331(5) definition. Under the Homeland Security Act of 2002.

[c] See https://www.fbi.gov/file-repository/fbi-dhs-domestic-terrorism-strategic-report-2023.pdf/view.

timeframe to post-9/11 Strategies (2001–2023). Other strategies and related documents considered by the committee are in Appendix A. In addition to evaluating documents, the committee was also briefed by representatives of various federal agencies through several information-gathering meetings (see full list of briefers Appendix A).

In conducting the strategies assessment, the committee has focused on chemical threats originating from non-state actors with or without state involvement (e.g., knowledge or capabilities sharing and other forms of support to enable chemical terrorism) but not the states themselves. The committee also conducted a high-level review of recent chemical terrorism events, leveraging the work of subject matter experts including members of the committee.

The recent rise of domestic terrorist events has motivated the committee to focus on this aspect along with foreign terrorism (whether directed abroad or at domestic targets). The study's emphasis on domestic terrorism is aligned with the 2023 Federal Bureau of Investigation (FBI) and Department of Homeland security (DHS) assessment and data report, which states:

> *"The threat posed by international and domestic threat actors has evolved significantly since 9/11. One of the most significant terrorism threats to the United States we face today is posed by lone actors and small groups of individuals who commit acts of violence motivated by a range of ideological beliefs and/or personal grievances. Of these actors, domestic violent extremists represent one of the most persistent threats to the United States today."* (FBI, 2023)

Furthermore, federal organizations that are key players in the chemical terrorism space recognize that the definition of domestic violent extremist (DVE) will need to be updated in order to accurately assess the current threat space:

> *"In 2021, the FBI, DHS I&A, and NCTC jointly updated the booklet, U.S. Violent Extremist Mobilization Indicators, which contains observable indicators to help bystanders or observers recognize behaviors that may indicate mobilization to violence. Unlike prior editions—which focused entirely on foreign terrorist-inspired, homegrown violent extremists—the 2021 edition was expanded to include indicators that apply across U.S.-based ideologically motivated violent extremists, including indicators validated as relevant for DVEs."*

These terms apply to terrorist groups that use a wide range of WMDs. Staying within the scope of this study, the committee placed more focus on chemical terrorism threats that were likely to cause immediate or significant impacts. Chemical terrorism threats considered include agents identified as chemical weapons as well as existing and emerging threats, including toxic industrial chemicals and materials (TICs and TIMs).

1.1.2 Committee's Approach

The committee recognized that its recommendations to improve strategies may only be implemented by actions requiring budget authority. The committee took a high-level view of this task and focused on identifying the most important technical, policy, and resource gaps with respect to identifying, preventing, countering, and responding to chemical threats and attacks.

The committee undertook an information-gathering strategy. As the study commenced, the committee collected policy documents regarding existing U.S. strategies against chemical weapons. As relevant strategies were published after the committee commenced, they were also evaluated. After the study committee had completed its data-gathering activities for this study, DoD issued the *2023 Department of Defense Strategy for Countering Weapons of Mass Destruction*. As a result, the study committee's report does not formally examine the new strategy document, but that strategy document is based on the *2022 National Defense Strategy*, which is discussed in the study committee's report. The study committee's ideas, findings, and recommendations still apply and merit full consideration.

Additionally, the committee received numerous briefings: from the study sponsor, organizations associated with strategies against chemical weapons, and subject-matter experts otherwise aligned with the statement of task (SOT). The briefings focused the committee on aspects of the broad chemical weapons landscape most and least aligned with the SOT. For budgetary assessment (Line 6 of the SOT), the committee compiled a series of national functions and the benefits of retaining these resources in future budgets.

Two parallel committees with similar charges are evaluating other national strategies against threats: one with a biological terrorism focus (*Assessing and Improving Strategies for Preventing, Countering, and Responding to Weapons of Mass Destruction Terrorism: Biological Threats*) and one with a nuclear terrorism focus (*Assessing and Improving Strategies for Preventing, Countering, and Responding to Weapons of Mass Destruction Terrorism: Nuclear Threat*). While radioactive substances are specific chemical elements (e.g., radon, isotopes of cobalt, cesium-137, polonium-210, uranium-235, plutonium-239, and americium-241), this chemical terrorism-focused committee decided that agents and terrorist activities where the harm primarily derives from radioactivity and the hazards derived therefrom are the purview of the nuclear committee.

Differentiation between the focus of the biological and chemical committees could be more difficult.[1] While some areas, such as bacteria (e.g. protein toxins pro-

---

[1] "Chemical terrorism" versus "biological terrorism" or "radiological terrorism" is not defined in the U.S. Code. "Chemical weapons" are defined in 18 U.S.C. 11B - CHEMICAL WEAPONS, see https://www.govinfo.gov/content/pkg/USCODE-2015-title18/html/USCODE-2015-title18-partI-chap11B.htm.

Chemical weapon—The term "chemical weapon" means the following, together or separately:

(A) A toxic chemical and its precursors, except where intended for a purpose not prohibited under this chapter as long as the type and quantity is consistent with such a purpose.

(B) A munition or device, specifically designed to cause death or other harm through toxic properties of those toxic chemicals specified in subparagraph (A), which would be released as a result of the employment of such munition or device.

(C) Any equipment specifically designed for use directly in connection with the employment of munitions or devices specified in subparagraph (B).

Purposes not prohibited by this chapter—The term "purposes not prohibited by this chapter" means the following: "

(A) Peaceful purposes.—Any peaceful purpose related to an industrial, agricultural, research, medical, or pharmaceutical activity or other activity.

(B) Protective purposes.—Any purpose directly related to protection against toxic chemicals and to protection against chemical weapons.

continued

duced by bacteria), viruses, and fungi are more easily delineated within the domain of biological terrorism; other materials like peptides and molecular toxins (e.g., cyanotoxin, anatoxin-a) blur this division (Fozo et al., 2008; Hayes, 2003). Chemical compounds and mixtures, including biologically derived molecules (peptides, prions, genetic material), were considered to be within the purview of this consensus study. The committee recognizes that this division does not rigorously define the chemical or biological terrorism boundary, however, they made this practical choice to impose limits on the scope of strategies to be investigated. The committee recognizes that this division is not without problems, especially in the context of increasingly interdisciplinary scientific approaches. This report will discuss threats at the intersection of biological and chemical terrorism between committees as it pertains to the charge.

1.2 CHEMICAL ENVIRONMENT

The next sections describe five trends that the committee considers to be significant—in varying ways and degrees—to the current and future environment of chemical threats: erosion of norms of nonuse of chemical agents (1.2.1); return to Great Power Competition (GPC)(1.2.2); emerging technologies (1.2.3); challenges to domestic capacity to respond (1.2.4); and threat of pharmaceutical-based agents (PBAs) (1.2.5). The committee's analysis, findings, conclusions, and recommendations intersect with these phenomena.

1.2.1 Erosion of Norms of Nonuse of Chemical Agents

While chemical agents have a long history, the world saw a transformative change in the scope and scale of their use on the battlefield in WWI continuing through the ongoing Syrian civil war. Recent use by authoritarian states to target defectors and dissidents has also revitalized international and domestic attention and interest in chemical weapons.

International law, laws of armed conflict, arms control, and other treaties, contribute to global norm formation (Brunnée, 2019; Deitelhoff, 2019; Katzenstein, 1996; Nyarko, 2018; Price, 1995; Tannenwald, 1999;). How and to what extent those legal frameworks apply to non-state actors, including VEOs is debated (Birdsall, 2016; Federer, 2019; O'Donnell, 2006; Wunderlich, 2020), including in the U.S. Supreme Court.[2] However, under the United Nations Security Council Resolution 1540 (UNSCR1540) countries are required to prevent terrorist access to WMD. Similarly, the Chemical Weapons Convention (CWC) requires adhering states to ensure that chemical weapons are not used within their territory.

(C) Unrelated military purposes.—Any military purpose of the United States that is not connected with the use of a chemical weapon or that is not dependent on the use of the toxic or poisonous properties of the chemical weapon to cause death or other harm.

(D) Law enforcement purposes.—Any law enforcement purpose, including any domestic riot control purpose and including imposition of capital punishment."

[2] Rasul v Bush (03-334) 321 F.3d 1134, reversed and remanded. https://www.law.cornell.edu/supct/html/03-334.ZO.html

> [A]ll States, in accordance with their national procedures, shall adopt and enforce appropriate effective laws which prohibit any non-State actor to manufacture, acquire, possess, develop, transport, transfer or use nuclear, chemical or biological weapons and their means of delivery, in particular for terrorist purposes, as well as attempts to engage in any of the foregoing activities, participate in them as an accomplice, assist or finance them. [and] all States shall take and enforce effective measures to establish domestic controls to prevent the proliferation of nuclear, chemical, or biological weapons and their means of delivery, including by establishing appropriate controls over related materials . . . (UNSCR, 2004, Pg. 2).

Non-state actors are not party to international treaties. Acts of terrorism directed at civilians and other noncombatants violate the norms of the laws of armed conflict. Nonetheless, the existing legal frameworks and associated norms can serve as guides and models for thinking about the violent use of unconventional weapons, like chemical agents, against civilians or against uniformed service members in noncombatant situations.

Since WWI, efforts domestically and internationally have led to the creation of international institutions and the establishment of international law and norms intended to reduce and eliminate the horrors of chemical weapons. These efforts grew out of visceral experiences from the war, during which chemical agents were extensively used, especially in the Western operational theater of trench warfare. Following WWI, states negotiated the 1925 Protocol for the Prohibition of the Use in War of Asphyxiating, Poisonous or Other Gases, and of Bacteriological Methods of Warfare, more commonly known as the Geneva Protocol. The Geneva Protocol had a relatively narrow scope and prohibited the use of biological and chemical weapons in interstate conflict, but it did not prohibit production, stockpiling, or testing of either class of weapons. The United States did not ratify the Geneva Protocol until 1975 and did so in the context of the ratification of the Biological Weapons Convention (BWC).

The CWC, which prohibits stockpiling, production, testing, and use of chemical weapons during interstate conflict, entered into force in 1997. As part of the CWC, a stand-alone international body, the Organization for the Prohibition of Chemical Weapons (OPCW), was created to oversee the implementation of the treaty, including the demilitarization of declared stockpiles. As of late 2023, more than 190 states/parties are included in the Convention.

In 2008, the bipartisan Commission on the Prevention of Weapons of Mass Destruction Proliferation and Terrorism, commonly known as the Graham/Talent WMD Commission, issued its final report. Among its conclusions was: "Unless the world community acts decisively and with great urgency, it is more likely than not that a WMD will be used in a terrorist attack somewhere in the world by the end of 2013." [3]

The Graham/Talent WMD Commission further clarified the type of WMD to which they were referring, asserting that "terrorists are more likely to be able to obtain and use

---

[3] World at Risk, https://web.archive.org/web/20090130205134/http://documents.scribd.com/docs/15bq 1nrl9aerfu0yu9qd.pdf, p 15. The text of the report reads: "The Commission believes that unless the world community acts decisively and with great urgency, it is more likely than not that a weapon of mass destruction will be used in a terrorist attack somewhere in the world by the end of 2013. The Commission further believes that terrorists are more likely to be able to obtain and use a biological weapon than a nuclear weapon."

a biological weapon than a nuclear weapon." Few considerations were made regarding chemical weapons in the report. By 2013, the world witnessed a resurgence of the use of chemical weapons by state-based actors.[4] Those uses were deployed for targeted assassination of political rivals and persons seen as politically threatening to authoritarian regimes; or as part of an inter-state civil war (i.e., Islamic State of Iraq and the Levant (ISIL) in the Syrian civil war). These are not the types of uses of chemical weapons that drove much of the Cold War era thinking, which was heavily influenced by military use in WWI and significantly less use (but some terrible exceptions and much stockpiling) in WWII. The last decade has seen renewed use of chemical weapons in conflicts and by authoritarian states as a means to limit rivals and others seen as threatening. A more detailed discussion of the history and analysis of the use of chemical agents by non-state actors, and terrorists, is discussed in chapter 4.

1.2.2 Return to Great Power Competition

The October 2022 National Security Strategy (NSS) began by noting a strategic shift in the international security environment: "the post-Cold War era is definitively over and a competition is underway between the major powers to shape what comes next" (NSS, 2022a). Referred to as Great Power Competition (GPC) or less commonly, strategic competition, the United States has shifted its strategic posture over the last decade from an emphasis on countering VEOs to addressing challenges from interstate competition. The NSS asserts that "the most pressing strategic challenge facing our vision is from powers that layer authoritarian governance with a revisionist foreign policy" (NSS, 2022b).

This intensified competition with the People's Republic of China (PRC or China) and the Russian Federation (Russia) "has profoundly changed the conversation about U.S. defense issues" (O'Rourke, 2022). **The Global War on Terrorism (GWOT), counterterrorism efforts, and U.S. operations in the greater Middle East and SW Asia—which had been at the center of U.S. security policy following the terrorist attacks of September 11, 2001—have given way to a stronger focus on China and Russia.** This strategic shift, which began with the United States "pivot" or "re-balance" to Asia announced in 2011 (Clinton, 2011), is resulting in changes throughout the USG, especially in those offices and agencies whose mission is directly related to national and international security. Budgets are being reconsidered and changes to organizational structures, including force planning, are being made or contemplated (O'Rourke, 2022). The committee tried to be attentive to how this major strategic shift is or could potentially impact the United States' ability to prevent, counter, and respond to chemical terrorism threats.

National security strategies are issued and updated on a periodic basis by various USG agencies. The *2022 National Defense Strategy Data Sheet* (DoD, 2022) (and the full strategy released 10/27/22) details the United States' shift to GPC, which could inadvertently create gaps in U.S. preparedness for chemical terrorism threats: hence

---

[4] This observation was made previously in Kosal (2019).

a motivation for examining current strategies as directed in the SOT (Box 1-1). The committee elaborates on this trend and its future implications on U.S. national strategies and budget in Chapter 7.

1.2.3 Emerging Technologies

Historically, terrorists have overwhelmingly pursued conventional weapons (largely guns and bombs) and have not shown a proclivity to innovate in general (Hoffman, 1993, 2001). Nonetheless, terrorist organizations, notably al-Qa'ida, have shown a capacity to exploit expectations regarding terrorist behavior and operations (National Commission on Terrorist Attacks Upon the United States, 2004). Scholars have looked at the question of terrorist innovation, both empirically and more speculatively (Ackerman, 2016; Dolnik, 2007; Gill et al., 2013; Kosal, 2009; Logan et al., 2021; Lubrano, 2021; Ranstorp and Normark, 2015; Tennenbaum and Kosal, 2021; Tishler, 2018).

Settling on a single definition or time horizon for what qualifies as an emerging technology is debated. One description delineates what qualifies as an emerging technology via five key attributes that a technology must possess in order to qualify as an emerging technology: radical novelty, relatively fast growth, coherence, prominent impact, and uncertainty and ambiguity (Rotolo, 2015). The White House Office of Science and Technology Policy (OSTP) has promulgated a list of specific technologies that have been identified as relevant to security concerns in its Critical and Emerging Technologies List Update (The White House, 2022a), but it does not define what an emerging technology is, which is not uncommon to encounter in documents.

While an exhaustive chemical threats inventory was out of scope for the study, it was necessary for the committee to identify what chemical threats they would use to compare strategies. Additionally, the role of strategies in encouraging and enabling cooperation among U.S. agencies at multiple levels, as well as with allied nations, was included in the report. Discussions regarding the likelihood and feasibility of deploying emerging threats technologies by various actors are further discussed in Chapter 2. Nonetheless, concerns about non-state actors using U.S. ingenuity against the country, especially in the context of emerging technology, are perennial concerns.

1.2.4 Domestic Capacity to Respond

Public and expert concerns about the ability of the United States to respond effectively or adequately have been heightened during the initial response to the COVID-19 global pandemic (Deslatte, 2020; Goldstein and Wiedemann, 2020; Goldfinch et al., 202; Hamilton et al., 2021; Latkin et al., 2020; Pollard and Davis, 2022;) and the rise in domestic partisanship (Funk et al., 2020; Gadarian et al., 2020; Milligan, 2020; Roberts, 2020; Van Green and Tyson, 2020). These concerns are especially impactful in working across levels of government—for example, among cities, counties, states, tribal authorities, and the federal government—as trust in government correlates positively with effective response in emergency situations (Lau et al., 2020). **Confidence in government institutions "has been identified as a cornerstone of the political**

system, particularly in crises such as natural disasters, economic crises, or pandemics" (Han et al., 2021). An investigative report, led by the executive director of the 9/11 Commission, Philip Zelikow, found that "the leaders of the United States could not apply their country's vast assets effectively enough in practice" (*Washington Post*, 2023). Questions remain regarding the domestic capacity to respond. While most questions are beyond the scope of this committee, the scope and scale of the broader impact on public confidence is an aspect that the committee considered important enough to highlight as a trend with impacts on the United States' ability to identify, counter, and respond to chemical terrorism threats.

In the context of the broader threat of terrorism, the government agency, Europol, highlighted how COVID-19 and the perceived inability of governments to respond, including those outside the United States, has affected terrorist groups. They specifically note that "for those advocating extremist ideologies, the crisis has emerged as an opportunity to advance their narrative" (EUROPOL, 2022) of U.S. capacity and capability to respond to terrorist incidents, including those that employ traditional, improvised, or emerging chemical agents. The effectiveness of the National Response Framework and other strategies implicitly relies on a robust capacity to respond. Further implications of this trend in the context of other specific strategies are discussed in Chapter 6.

1.2.5 Pharmaceutical-based agents (PBAs)

In considering the intersections of pharmaceutical-based agents (PBAs) and the threat of chemical terrorism, the foremost reason for this consideration is that it may be a proxy for thinking and preparing to respond to other agents. This concept is not new; the world witnessed the potential for opioid-like compounds to cause significant fatalities in 2002 when Russian special police used an aerosolized fentanyl analog, carfentanil, to end the siege of a Moscow theater by Chechen separatists. That incident has prompted a good deal of writing on concerns related to riot-control agents (RCA) (Crowley, 2016; Fidler, 2005; Klotz et al., 2003; Martínková and Smetana, 2020; Robinson, 2007; Timperley et al., 2018). That event also points to concerns regarding an emerging class of agents, including their lethality and potential operational use.

PBAs that affect the central nervous system are a class of chemicals that are authorized and used for legitimate medical, veterinary, pharmaceutical, chemical production, agricultural, and other purposes (Caves and Carus, 2022). Developed as anesthetics (pain reducers or sensation reducers), analgesics (pain relievers), anticonvulsants, anorexiants (appetite suppressants), anti-Parkinson agents, cholinesterase inhibitors (nerve agent countermeasures), and calmatives (sedatives) (Daggett, 2007), they are a subset of incapacitating agents. If used in a contraindicated manner, in excess (overdose), or in certain exposure contexts, they can cause incapacitation, injury, or death. These substances include opioids, like fentanyl; but also nonsteroidal anti-inflammatory drugs (NSAIDs), like ibuprofen; barbiturates and benzodiazepines,[5] used to treat sei-

---

[5] For example, the opioid benzodiazepine, Seizalam, was cited as an example of countermeasure for chemical terrorism during one of the committee's briefings.

zures; and carfentanil, used legitimately in large animal veterinary practices; or less so by Russian Spetsnaz (Special Police) Forces in 2002 in response to Russian domestic terrorists who occupied a crowded theater, taking 850 hostages (also known as the 2002 Nord-Ost siege) (Riches et al., 2012).

Opioids are not inherently chemical weapons.[6] Designation of something as a chemical weapon is not solely based on its toxicity (see definition of chemical weapon Table 1-1). Opioids have legitimate and beneficial uses. (World Health Organization, n.d.) Local or criminally motivated attacks are not chemical terrorism.[7] The committee notes that a careful unpacking of the issues surrounding opioids is necessary due to the complexity and need to clearly state that the current epidemic (The White House, 2022b)[8] is not chemical terrorism and that pharmaceutical pain medications, licit or illicitly obtained, are not chemical weapons. The U.S. opioid epidemic may, however, be an example for thinking about a number of pressing issues directly related to U.S. strategies and efforts to reduce the threat of chemical weapons by both state and non-state actors in the twenty-first century.

1.3 REPORT ORGANIZATION

The remainder of this report discusses the trends mentioned in section 1.2 and the current state of U.S. strategies for identifying, preventing/countering, and responding to chemical terrorism. Chapter 2 introduces the chemical threat landscape—both baseline and emerging threats that set the stage for the committee's analysis. Then, a systematic methodology to assess the adequacy of USG strategies is presented in Chapter 3. The subsequent chapters provide an assessment of the strategies from the perspective of identifying chemical threats (Chapter 4), preventing or countering chemical attacks (Chapter 5), and responding to chemical attacks or chemical hazards (Chapter 6). Throughout the report, technical, policy, and resource gaps in the strategies are discussed. Findings, conclusions, and recommendations related to each framework (identify, prevent/counter, or respond) are presented. A brief summary can be found at the end of each chapter. Finally, Chapter 7 covers major themes that cut across the previous chapters and provides broader recommendations applied to the overall national strategy (i.e., shift toward GPC) beyond the documents assessed by the committee.

REFERENCES

Ackerman, Gary A. 2016. "Comparative Analysis of VNSA Complex Engineering Efforts." *Journal of Strategic Security* 9(1): 119–133. JSTOR. http://www.jstor.org/stable/26465417.
Birdsall, Andrea. 2016. "But We Don't Call It 'torture'! Norm Contestation During the U.S. 'War on Terror.'" *International Politics* 53(2):176–197.

---

[6] See 18 U.S.C. 11B - CHEMICAL WEAPONS, see https://www.govinfo.gov/content/pkg/USCODE-2015-title18/html/USCODE-2015-title18-partI-chap11B.htm.

[7] See https://www.law.cornell.edu/supremecourt/text/12-158.

[8] In-line with the statement of task, the committee primarily focused on non-state actors and did not assess U.S. strategies to address threats or attacks from state actors.

Brunnée, Jutta, and Stephen J. Toope. 2019. "Norm Robustness and Contestation in International Law: Self-Defense against Non-State Actors." *Journal of Global Security Studies* 4(1):73–87.

Latkin, Carl A., L. Dayton, J. C. Strickland, B. Colon, R. Rimal, and B. Boodram. 2020. "An Assessment of the Rapid Decline of Trust in U.S. Sources of Public Information about COVID-19." *Journal of Health Communication* 25(10): 764–773. doi: 10.1080/10810730.2020.1865487.

Caves, Jr., John P., and W. Seth Carus. 2022. "Controlling Chemical Weapons in the New International Order." *Proceedings NDU CSWMD*, August 2022. https://wmdcenter.ndu.edu/

Clinton, H. 2011. America's Pacific Century. Foreign Policy, 11 October 2011. https://foreign-policy.com/2011/10/11/americas-pacific-century/& "Barack Obama says Asia-Pacific is 'top U.S. priority,'" BBC, 17 November 2011. https://www.bbc.com/news/world-asia-15715446.

Crowley, M. 2016. *Chemical Control: Regulation of Incapacitating Chemical Agent Weapons, Riot Control Agents and Their Means of Delivery*. Basingstoke: Palgrave Macmillan.

Daggett, M. 2007. "Pharmacological Overview of Calmatives." NCJ No. 220986, National Criminal Justice Reference Service (NCJRS), Dept. of Justice. https://www.ojp.gov/ncjrs/virtual-library/abstracts/pharmacological-overview-calmatives.

Deitelhoff, N., L. Zimmermann. 2019. "Norms under Challenge: Unpacking the Dynamics of Norm Robustness." *Journal of Global Security Studies* 4(1): 2–17. https://doi.org/10.1093/jogss/ogy041.

Deslatte, Aaron. 2020. "The Erosion of Trust during a Global Pandemic and How Public Administrators Should Counter It." *The American Review of Public Administration* 50(6-7): 489–496. https://doi.org/10.1177/0275074020941676.

DHS (Department of Homeland Security). 2017. Risk Lexicon. https://www.dhs.gov/sites/default/files/publications/18_0116_MGMT_DHS-Lexicon.pdf.

DoD (U.S. Department of Defense). 2022. "2022 National Defense Strategy of the United States of America." https://media.defense.gov/2022/Oct/27/2003103845/-1/-1/1/2022-national-defense-strategy-npr-mdr.pdf.

DoD. 2021. Dictionary. https://www.jcs.mil/Portals/36/Documents/Doctrine/pubs/dictionary.pdf?ver= J3xmdacJe_L_DMvIUhE7gA%3d%3d.

Dolnik, A. 2007. *Understanding Terrorist Innovation: Technology, Tactics and Global Trends* (1st ed.). Routledge. https://doi.org/10.4324/9780203088944.0

EUROPOL. 2022. European Union Terrorism Situation and Trend Report. https://www.europol.europa.eu/cms/sites/default/files/documents/Tesat_Report_2022_0.pdf.

FBI (Federal Bureau of Investigation). 2023. https://www.fbi.gov/file-repository/fbi-dhs-domestic-terrorism-strategic-report-2023.pdf/view.

Fidler, D. P. 2005. "The Meaning of Moscow: 'Non-lethal' Weapons and International Law in the Early Twenty-first Century." *International Review of the Red Cross* 87(859): 531.

Fozo, E. M., M. R. Hemm, G. Storz. 2008. "Small Toxic Proteins and the Antisense RNAs That Repress Them." *Microbiol Mol Biol Rev*. 72(4): 579–89. https://www.ncbi.nlm.nih.gov/pmc/articles/PMC2593563.

Funk, C., B. Kennedy, C. Johnson. 2020. "Trust in Medical Scientists Has Grown in the U.S., But Mainly among Democrats." Pew Research Center Science & Society. https://www.pewresearch.org/science/2020/05/21/trust-in-medical-scientists-has-grown-in-u-s-but-mainly-among-democrats.

Gadarian, Shana Kushner, Sara Wallace Goodman, Thomas B. Pepinsky. 2020. "Partisanship, Health Behavior, and Policy Attitudes in the Early Stages of the COVID-19 Pandemic." (March 27, 2020). http://dx.doi.org/10.2139/ssrn.3562796.

Gill, P., J. Horgan, S. T. Hunter, and L. D. Cushenbery. 2013. "Malevolent Creativity in Terrorist Organizations." *J Creat Behav* 47:125–151. https://doi.org/10.1002/jocb.28.

Goldfinch, Shaun, Ross Taplin, and Robin Gauld. 2021. "Trust in Government Increased during the COVID-19 Pandemic in Australia and New Zealand." *Austr. J. Pub. Admin.* 80(1): 3–11. https://doi.org/10.1111/1467-8500.12459.

Goldstein, D. A. N., J. Wiedemann. 2020. "Who Do You Trust? The Consequences of Political and Social Trust for Public Responsiveness to COVID-19 Orders." (SSRN Scholarly Paper ID 3580547). Social Science Research Network. https://doi.org/10.2139/ssrn.3580547.

Hamilton, L. C., and T. G. Safford. 2021. "Elite Cues and the Rapid Decline in Trust in Science Agencies on COVID-19." *Sociological Perspectives* 64(5): 988–1011. doi:10.1177/07311214211022391.

Han, Q., B. Zheng, M. Cristea, M. Agostini, J. Bélanger, B. Gützkow, N. Leander. 2021. "Trust in Government Regarding COVID-19 and Its Associations with Preventive Health Behaviour and Prosocial Behaviour during the Pandemic: A Cross-sectional and Longitudinal Study." *Psychological Medicine*, 1–11. doi:10.1017/S0033291721001306.

Hayes, F. 2003. "Toxins-antitoxins: Plasmid Maintenance, Programmed Cell Death, and Cell Cycle Arrest." *Science* 301: 1496–1499.

Hoffman, Bruce. 1993. "Terrorist Targeting: Tactics, Trends, and Potentialities." *Terrorism and Political Violence* 5(2): 12–29. doi: 10.1080/09546559308427205, https://www.tandfonline.com/doi/abs/10.1080/09546559308427205.

Hoffman, Bruce. 2001. "Change and Continuity in Terrorism." *Studies in Conflict and Terrorism.* 24(5): 417–428. https://doi.org/10.1080/105761001750434268.

Federer, J. P. 2019. "We Do Negotiate with Terrorists: Navigating Liberal and Illiberal Norms in Peace Mediation." *Critical Studies on Terrorism* 12(1): 19–39. doi: 10.1080/17539153.2018.1472727.

Katzenstein P. J., ed. 1996. *The Culture of National Security: Norms and Identity in World Politics.* New York: Columbia University Press.

Klotz, L., M. Furmanski, and M. Wheelis. 2003. "Beware the Siren's Song: Why 'Non-Lethal' Incapacitating Chemical Agents Are Lethal." https://fas.org/programs/bio/chemweapons/documents/sirens_song.pdf.

Kosal, M. 2019. H-Diplo/ISSF Article Review 114, "The Future of Chemical Weapons: Implications from the Syrian Civil War," March 2019. https://networks.hnet.org/node/28443/discussions/4084042/h-diploissf-article-review-114-kosal-%E2%80%9C-future-chemical-weapons.

Kosal, M. E. 2009. *Nanotechnology for Chemical and Biological Defense.* Springer Academic Publishers: New York. June 2009. http://www.springer.com/us/book/9781441900616.

Lau, L. S., G. Samari, R. T. Moresky, S. E. Casey, S. P. Kachur, L. F. Roberts, and M. Zard. 2020. "COVID-19 in Humanitarian Settings and Lessons Learned from Past Epidemics." *Nature Medicine* 26(5): 647–648. https://doi.org/10.1038/s41591-020-0851-2.

Logan, M. K., A. Damadzic, K. Medeiros, G. S. Ligon, and D. C. Derrick. 2021. "Constraints to Malevolent Innovation in Terrorist Attacks." *Psychology of Aesthetics, Creativity, and the Arts.* https://doi.org/10.1037/aca0000385.

Lubrano, M. 2021. "Navigating Terrorist Innovation: A Proposal for a Conceptual Framework on How Terrorists Innovate." *Terrorism and Political Violence* 35(2). https://www.tandfonline.com/doi/full/10.1080/09546553.2021.1903440.

Martínková, H., and M. Smetana. 2020. "Dynamics of Norm Contestation in the Chemical Weapons Convention: The Case of 'Non-Lethal Agents.'" *Politics* 40(4): 428–443. https://doi.org/10.1177/0263395720904605.

Milligan, S. 2020. "The Political Divide over the Coronavirus." *U.S. News & World Report.* https://www.usnews.com/news/politics/articles/2020-03-18/the-political-divide-over-the-coronavirus.

National Commission on Terrorist Attacks Upon the United States. 2004. The 9/11 Commission Report: Final Report of the 9/11 Report.

NSS. 2022a. https://www.whitehouse.gov/wp-content/uploads/2022/10/Biden-Harris-Administrations-National-Security-Strategy-10.2022.pdf, 6.

NSS. 2022b. https://www.whitehouse.gov/wp-content/uploads/2022/10/Biden-Harris-Administrations-National-Security-Strategy-10.2022.pdf, 8.

Nyarko, J. 2018. "Giving the Treaty a Purpose: Comparing the Durability of Treaties and Executive Agreements." *The American Journal of International Law* 113(1): 54–89.

O'Rourke, R. 2022. "Great Power Competition: Implications for Defense—Issues for Congress." Congressional Research Service, R43838, November 08, 2022, 2. https://crsreports.congress.gov/product/details?prodcode=R43838.

O'Donnell, D. 2006. "International Treaties against Terrorism and the Use of Terrorism during Armed Conflict and by Armed Forces." *International Review of the Red Cross* 88(864): 853–880. doi:10.1017/S1816383107000847.

Pollard, M. S., and L. M. Davis. 2022. "Decline in Trust in the Centers for Disease Control and Prevention During the COVID-19 Pandemic." *Rand Health Q* 9(3): 23.

Price, R. 1995. "A Genealogy of the Chemical Weapons Taboo." *International Organization* 49(1): 73–103.

Ranstorp, M., and M. Normark (Eds.). 2015. *Understanding Terrorism Innovation and Learning: Al-Qaeda and Beyond* (1st ed.). Routledge. https://doi.org/10.4324/9781315726816.

Riches, J. R., R. W. Read, R. M. Black, N. J. Cooper, and C. M. Timperley. 2012. "Analysis of Clothing and Urine from Moscow Theatre Siege Casualties Reveals Carfentanil and Remifentanil Use." *Journal of Analytical Toxicology* 36: 647–656.

Roberts, D. 2020. "Partisanship is the Strongest Predictor of Coronavirus Response." *Vox.* https://www.vox.com/science-and-health/2020/3/31/21199271/coronavirus-in-us-trump-republicans-democrats-survey-epistemic-crisis.

Robinson, J. P. P. 2007. "Non-Lethal Warfare and the Chemical Weapons Convention." http://www.sussex.ac.uk/Units/spru/hsp/Papers/421rev3.pdf.

Rotolo, Daniele, Diana Hicks, and Ben R. Martin. 2015. "What Is an Emerging Technology?" *Research Policy* 44(10): 1827–1843.

Tannenwald, N. 1999. "The Nuclear Taboo: The United States and the Normative Basis of Nuclear Non-Use." *International Organization* 5 (3): 433–468.

Tennenbaum, M., and M. E. Kosal. 2021. "The Interplay Between Frugal Science and Chemical and Biological Weapons: Investigating the Proliferation Risks of Technology Intended for Humanitarian, Disaster Response, and International Development Efforts," in *Weapons Technology Proliferation: Diplomatic, Information, Military, Economic Approaches to Proliferation*, 153–204. https://www.springer.com/us/book/9783030736545.

Timperley, C. M. 2018. J. E. Forman, P. Åas, M. Abdollahi, D. Benachour, A. S. Al-Amri, A. Baulig, R. Becker-Arnold, V. Borrett, F. A. Cariño, C. Curty, D. Gonzalez, M. Geist, W. Kane, Z. Kovarik, R. Martínez-Álvarez, R. Mikulak, N. M. F. Mourão, S. Neffe, E. DS. Nogueira, P. Ramasami, S. K. Raza, V. Rubaylo, A. E. M. Saeed, K. Takeuchi, C. Tang, F. Trifirò, F. M. van Straten, A. G. Suárez, F. Waqar, P. S. Vanninen, M. Zafar-Uz-Zaman, S. Vučinić, V. Zaitsev, M. S. Zina, S. Holen, and F. Nurul Izzati. "Advice from the Scientific Advisory Board of the Organisation for the Prohibition of Chemical Weapons on Riot Control Agents in Connection to the Chemical Weapons Convention. *RSC Adv.*, 8:41731–41739. https://pubs.rsc.org/en/content/articlehtml/2018/ra/c8ra08273a.

Tishler, Nicole A. 2018. "Trends in Terrorists' Weapons Adoption and the Study Thereof." *International Studies Review* 20(3): 368–394. https://doi.org/10.1093/isr/vix038.

United Nations Security Council Resolution (UNSCR). 2004. Resolution 1540. https://documents-dds-ny.un.org/doc/UNDOC/GEN/N04/328/43/PDF/N0432843.pdf?OpenElement.

Van Green, T., and A. Tyson. 2020. "5 Facts about Partisan Reactions to COVID-19" in the *U.S. Pew Research Center*. https://www.pewresearch.org/fact-tank/2020/04/02/5-facts-about-partisan-reactions-to-covid-19-in-the-u-s.

Washington Post. 2023. https://www.washingtonpost.com/opinions/2023/04/24/covid-pandemic-government-response-report. https://covidcrisisgroup.org.

Weapons of Mass Effect Task Force. 2006. "Preventing the Entry of Weapons of Mass Effect into the United States." Homeland Security Advisory Council. https://www.dhs.gov/xlibrary/assets/hsac_wme-report_20060110.pdf.

The White House. 2012. Remarks by the President to the White House Press Corps, The White House, August 20, 2012. https://obamawhitehouse.archives.gov/the-press-office/2012/08/20/remarks-president-white-house-press-corps.

The White House. 2022a. https://www.whitehouse.gov/wp-content/uploads/2022/02/02-2022-Critical-and-Emerging-Technologies-List-Update.pdf.

The White House. 2022b. National Drug Control Strategy, The White House Executive Office of the President. Office of National Drug Control Policy. https://www.whitehouse.gov/wp-content/uploads/2022/04/National-Drug-Control-2022Strategy.pdf.

World Health Organization (WHO). (n.d.). "Essential Medicines Lists." https://www.who.int/groups/expert-committee-on-selection-and-use-of-essential-medicines/essential-medicines-lists.

Wunderlich, Carmen. 2020. *Rogue States as Norm Entrepreneurs*. Springer Academic Publishers. https://link.springer.com/book/10.1007/978-3-030-27990-5.

2

Chemical Threats and U.S. Governmental and Nongovernmental Institutions That Play a Role (The Threat and the Who's Who)

The following sections provide a broad overview of the chemical threat landscape, including social considerations, range of baseline threats, and their characterizations. Brief descriptions of chemical agent delivery methods are also provided. Later, a discussion around emerging technologies and their role in the threat landscape as well as key actors involved in this space is presented. The intent of chapter 2 is to lay out the different considerations involved in assessing this complex setting. If the reader prefers to omit this preliminary information, they should proceed directly to chapter 3, to learn about the assessment methodology applied in this study.

2.1 COMPLEX CHEMICAL THREAT LANDSCAPE

Terrorism involving chemical, biological, radiological, or nuclear (CBRN) agents has been extremely rare within the annals of nonstate terrorism overall (START, 2022). However, chemical terrorism has been the most common—and successful in terms of casualties caused—form of CBRN terrorism to date. According to the Profiles of Incidents involving CBRN and Nonstate Actors (POICN) Database (Binder and Ackerman, 2019), terrorist interest in, pursuit, and use of chemical weapons constitutes ~ 76 percent of all CBRN terrorism, with ~ 400 incidents of ideologically motivated actors (Binder et al., 2017)[1] pursuing chemical weapons (agent + delivery system) recorded between 1990–2020.[2] Approximately 50 percent of these incidents have resulted in the actual use of an agent (Binder and Ackerman, 2020) and at least half of the 400 incidents involv-

---

[1] POICN does not capture purely criminal uses with no ideological component, but there are estimated to be even more of these, such as poisonings of business rivals.

[2] There were 11 additional cases of adversary interest in CW that did not reach the level of a defined plot, but indicated actions that might lay the groundwork for an actual plot. Examples of such "protoplots" include discovery of a chemical weapons manual or hiring a scientist with a weapons' specialty (definition of a protoplot in Binder et. al. (2017) POICN *Database Codebook Version 8.71* [National Consortium for the Study of Terrorism and Responses to Terrorism: College Park, Maryland]).

ing chemical agents have been assessed as being of interest to those concerned about mass-casualty terrorism.[3] An analysis of the Global Terrorism Database, which covers a longer timespan than POICN (1970–2015) but only includes actual uses of chemical agents, counts 292 chemical terrorism incidents (Santos et. al., 2019).

These incidents of chemical terrorism and attempted terrorism have involved 100 different perpetrators motivated by different ideologies. For the period between 1990 and 2020, the geographic distribution of countries where chemical terrorism incidents have occurred is extensive (see Figure 2-1). This includes 68 cases in the United States, with 36 uses of a chemical agent by perpetrators in the United States. Chemical terrorism has also utilized an array of chemical agents, encompassing many commonly available (often referred to as "low-end") agents and several toxic industrial chemicals (TICs) and toxic industrial materials (TIMs), but also including some chemical agents that have traditionally been developed in the military context (see Table 2-1).

Specific mention of the threat from TICs and TIMs is warranted, since these agents have historically accounted for a large fraction of terrorist incidents that involve chemicals. According to the POICN Database (see Table 2-1), at least 90 of the roughly 200 uses of chemical agents by terrorists involved TICs or TIMS, whereas in the data-

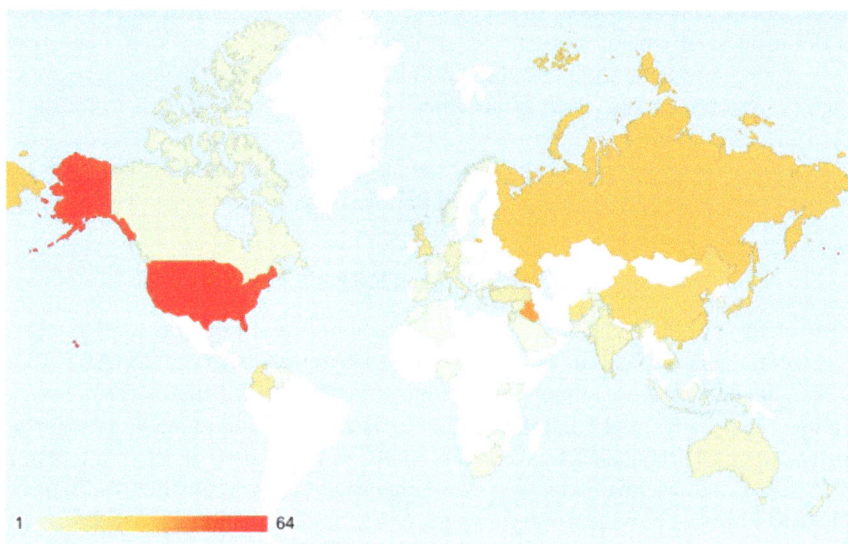

FIGURE 2-1 Geographic distribution of actual or intended target countries of chemical terrorism events (where known) recorded from 1990 to 2020. Terrorism events include plots, attempted acquisitions, possession of agents or weapons, and actual uses.
SOURCE: POICN Database (Binder and Ackerman, 2020).

---

[3] These are events which the POICN Database coded as "Heightened Interest," which denotes involvement of at least five total casualties, a CBRN agent classed as a warfare agent, fissile materials, or having at least moderately sophisticated agent weaponization (Binder et. al., 2017).

set developed by Santos and colleagues (2019), the category of "corrosives," which includes chlorine, was listed as the most commonly used chemical agent in an attack, and cyanide compounds were also used frequently. They also found that the lethality of chemical attacks using TICs was significantly lower compared to the lethality of attacks using nerve agents. However, the ubiquity and large volumes of TICs/TIMs mean that the scope for possible harm is substantial. It has been estimated that in the United States, there were 123 facilities that possessed sufficient quantities of TICs/TIMs capable of killing one million or more people (Kosal, 2006).

Additionally, the Government Accountability Office (GAO) reported that facilities storing or manufacturing hazardous chemicals could be targeted by terrorists; the salient example cited is the use of chlorine in Syria in 2018 (GAO, 2020). This warning was reiterated by scholars from the Center for the Study of Weapons of Mass Destruction at National Intelligence University (NIU) who noted that over 90 percent of the chemical weapon attacks in the Syrian civil war involved TICs, principally chlorine (Caves and Carus, 2021). Historically, other works have substantiated the risk. A 2006 Congressio-

TABLE 2-1 Agents Involved in Chemical Terrorism Incidents [a]

Chemical	Primary Use	# of Incidents of Interest, Pursuit or Use	# of Incidents of Use
Hydrogen Cyanide	TIC	47	14
Chlorine	Military / TIC	41	27
Butyric Acid	IC	24	24
Sodium Cyanide	TIC	23	2
Sarin	Military	20	4
Mustard Agent	Military	15	6
Potassium Cyanide	TIC	14	2
VX	Military	14	9
Ammonia Compounds	TIC	10	7
Unspecified Cyanide Salt	TIC	10	1
Arsenic	TIC	8	3
Hydrochloric Acid	TIC	6	1
Lachrymatory Acid/Pepper Spray/Mace	LE	6	5
CS Gas	LE	5	4
Nitric Acid	TIC	5	0
Sulfuric Acid	TIC	5	1

[a] Primary use listed include toxic industrial chemicals (TIC), military, law enforcement (LE), and industrial chemical (IC).
NOTES: *Others (Use Cases Bolded):* **mercury, strychnine, mercuric chloride,** nicotine sulfate, **sodium hydroxide, acetone, benzene,** dimethyl sulfoxide, halothane, hydrogen fluoride, malathion, **methanol, phosgene, sodium hypochlorite,** sodium monofluoroacetate, **phosphine (PH$_3$), aldrin,** atropine, brodifacoum, BZ, carbofuran, chloroform, chlorophenyl silatrane, chloropicrin, digoxin, diisopropyl fluorophosphate, **Drano, endrin, hydrazine,** ketamine, **lewisite,** methomyl, **methylene blue,** paraquat, **peniprazine chloride,** phenol, sodium chlorate, sulfur, tabun, **tellurium,** tetraethylammonium bromide (TEAB), vinegar, warfarin, **cyanic acid, vinyl trichlorosilane,** hydrogen sulfide, sodium chlorate.
SOURCE: POICN Database (Binder and Ackerman, 2020).

nal Research Service report presented on the threat posed by terrorists opportunistically employing TICs and TIMs in a chemical attack (CRS, 2006), noting specifically that chemical facilities might be targets of opportunity for terrorists to release chemicals into communities. The report also suggested that this risk was increasing, with the possibility of severe consequences on human health and the environment. Casillas and colleagues assessed that the availability or access to TICs/TIMs makes their use in terrorist activities more likely because they are not tightly controlled like chemical warfare agents (Casillas et al., 2021). The *2018 National Strategy for Countering WMD and Terrorism* (GovInfo, 2018) also mentions the use of TICs as an active threat, and it recommends tightened security practices for academic and industrial sectors.

For the past forty years, 21 reported attacks directed toward chemical facilities have been identified with seven incidents of terrorism directed at individuals. (Kosal, 2007). While this is a small number compared to general attacks, a large magnitude of casualties could occur following a well-organized attack on chemical-based facilities.

While the number of chemical terrorism incidents has risen and fallen over time, there is no empirical or analytical indication that the threat is disappearing (see Figure 2-2), especially with several incidents within the past two decades of terrorists using or pursuing various chemical agents, including those classed as warfare agents.

To obtain a better understanding of the baseline chemical terrorism threat presented above, it is necessary to parse the threat into the two basic components of motivation (incentives and disincentives) and capability.

2.1.1 Incentives and Disincentives for Using a Chemical Agent or Weapon

While a complete treatment of terrorist CBRN motivations is beyond the scope of this assessment, the reasons why some terrorists and not others pursue chemical weapons are essential to explore in at least some depth. At the outset, it is worth noting that to arrive at the decision to pursue a chemical weapon, in most cases a terrorist needs to make several specific choices (even if these are done implicitly). Taking a terrorist actor's general desire to employ some weapon or violent tactic to achieve its goals as a starting point, the first

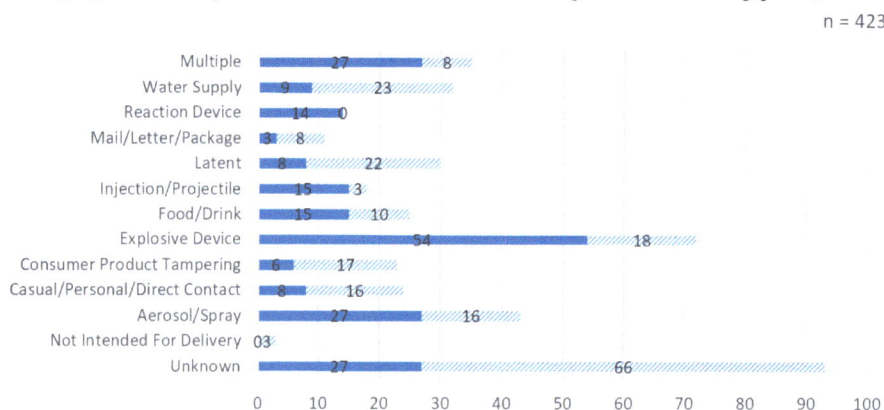

FIGURE 2-2 Chemical terrorism incidents by (intended) delivery method.
SOURCE: POICN Database (Binder and Ackerman, 2020).

choice is whether to use "conventional" terrorist modalities like guns and explosives or to innovate by using a novel or unconventional attack method. In only a relatively small proportion of circumstances will a terrorist actor seek to innovate in any way (Cameron, 1999; Dolnik, 2007; Hoffman, 1992; Jenkins, 1986), in which case they may decide to pursue a novel tactic combined with a conventional weapon (such as flying an airplane into a building as on September 11, 2001), new organizational approaches, or to pursue an unconventional weapon. For the subset of actors seeking to pursue an unconventional weapon, the choice is often between a chemical weapon and other types of unconventional weapon (e.g., biological, electromagnetic, or radiological). It is important to recognize that at each of these stages incentives and disincentives, opportunities, and obstacles exist. These factors are considered by the terrorist decision-maker, with only a relatively limited subset of pathways leading to the final decision to pursue a chemical weapon.

Incentives that may attract terrorists to unconventional weapons in general, including chemical weapons, followed by examples:

- *Strategic or Operational Advantages*—causing massive numbers of casualties through punishment or revenge; exerting a disproportionate psychological impact on the target society; gaining extensive publicity; deterrence; provoking government backlash; or forcing increased spending on defenses.
- *Tactical Advantages*—conducting a covert attack and achieving area contamination.
- *Organizational Benefits*—building status to assist in leadership struggles or intergroup rivalries and diversifying the weapons portfolio.
- *Ideology*—emulating sacred texts or myths; "techno-fetishism"; apocalyptic purification through sacrificial acts (Lifton, 2007).

Disincentives that could explain the relative rarity of unconventional weapon attacks by terrorists include the following with examples:

- *Status quo inertia*—unconventional weapons are usually neither necessary nor appropriate for achieving the terrorist's goals; risk aversion and lack of innovativeness; perception that developing a CBRN capability is too difficult or creates too much of an opportunity cost; the terrorist's operational tempo, strategic time horizon, and sense of urgency preclude the development of a complex weapons capability.
- *Negative consequences*—fear of reprisal by the targeted parties or the international community from the use of a banned weapon; concern about the loss of constituency support from the use of "morally illegitimate" weapons.
- *Ideological proscription*—where the effects of the unconventional weapon are anathema to the actor's ideology for a variety of reasons.
- *Fear of self-harm*—concern regarding the safety hazards of working with many unconventional weapons agents and precursors.

Terrorist actors for whom one or more of the above incentives operate strongly and for whom the disincentives are less salient are most likely to select unconventional over conventional weapons. Within this relatively small subset of actors, there are a few factors that would push a terrorist particularly toward chemical weapons as opposed to other unconventional weapons. Chief among these is the perception that chemical weapons are easier to acquire and deploy than other weapon types, while still satisfying the terrorist's strategic, operational, tactical, or organizational objectives. Two attributes that are often associated with a greater likelihood of unconventional weapon selection are a) having a religious ideology, since a "divine imprimatur" provides a basis for overcoming the ethical and social barriers to using these weapons, and b) cult-like organizations with charismatic leaders (see Bale, 2017; Bale and Ackerman, 2009). While quantitative studies have yielded mixed results for the relevance of these factors, they do appear to feature prominently in the chemical weapons attacks with the highest numbers of casualties.

Beyond general CBRN selection, other factors that could influence the pursuit of chemical weapons are shown in Table 2-2. The information in the table draws primarily from a previous study on the psychology of chemical and biological nonstate adversaries. (Ackerman et. al. 2017a; Binder et. al. 2017).

The above discussion provides context for understanding the empirical record of those who have pursued or used chemical weapons. Figures 2-3 and 2-4 show, respectively, the perpetrator type (e.g., religious extremist groups, ethnonationalist groups, political groups, and others) and the general motivations behind recorded cases of chemical weapons pursuit by terrorists. Additionally, Table 2-3 enumerates the types of entities involved. Furthermore, Tables 2-4, 2-5, and 2-6 provide lists of formal terrorist groups who have pursued a chemical weapons capability most prolifically.

TABLE 2-2 Factors Influencing Chemicals Relative to Other Unconventional Weapons

Factor	Influence on Chemical Weapons Selection
Perceived availability of chemical agents or precursors (including stored chemicals) relative to other weapon types	+ +
Leadership or operational cadre have a background in chemistry or the chemical industry	+
Proclivities (e.g., fetish) by leaders or operational commanders specifically toward chemicals	+++
Ideological drivers specifically involving CW	+++
Perceived prior use of CW against actors or constituents (revenge motive)	+++
Ideological proscription of chemical weapons or the effects thereof	-
Constituency intolerance of chemical weapons or the effects thereof[a]	- -
Rejection of modern technology	-
Leadership aversion to chemicals or safety concerns	-

[a] Merely having a constituency is not sufficient to dissuade a terrorist from selecting chemical weapons, as seen by the many ethnonationalists and other secular groups with constituencies who have pursued chemical weapons (see Figure 2-3).

NOTES: The symbols reflect a rough order-of-magnitude estimate of the degree to which each factor influences chemical weapons selection, with a (+) indicating a positive influence on the decision to select chemical weapons out of CBRN, a (-) indicating a negative influence and the number of symbols indicating the relative magnitude of the influence (low, moderate, or strong)

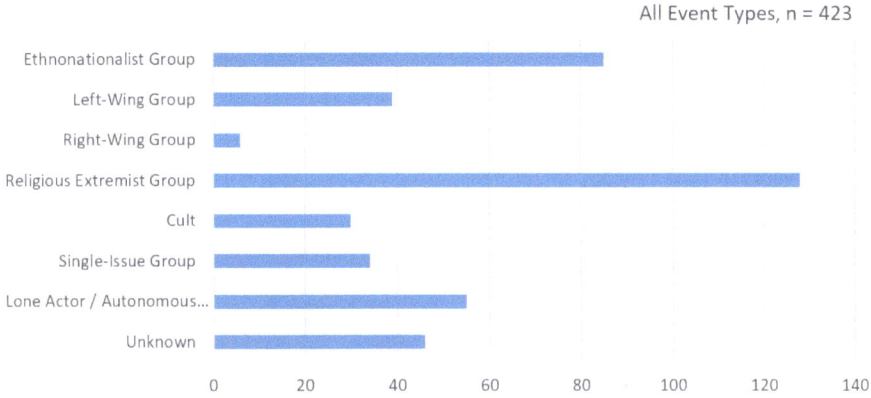

FIGURE 2-3 Interest in or pursuit of chemical weapons by perpetrator.
SOURCE: POICN Database (Binder and Ackerman, 2020).

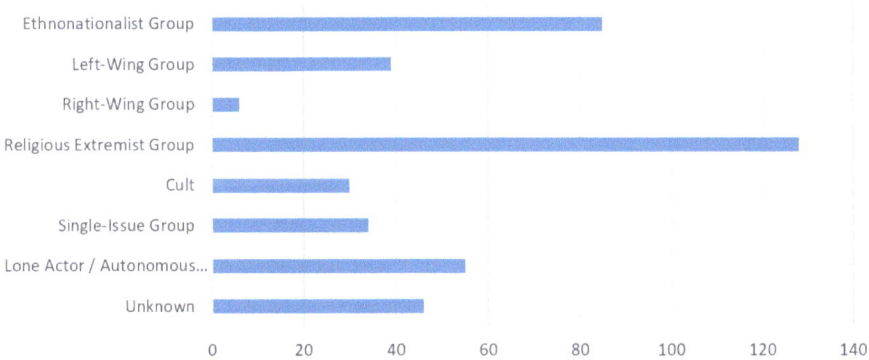

FIGURE 2-4 Interest in or pursuit of chemical weapons by general objective of perpetrator.
SOURCE: POICN Database (Binder and Ackerman, 2020).

TABLE 2-3 Number of Different Perpetrator Entities

Entity Type	All Incidents [% of Known Perpetrators]	Heightened Interest Incidents Only [% of Known Perpetrators]	Use Cases Only [% of Known Perpetrators]
Formal / Identified Organizations	78 [58%]	38 [59%]	28 [67%]
Unnamed / Unaffiliated Cells	18 [13%]	10 [16%]	6 [19%]
Individuals	38 [28%]	16 [25%]	8 [14%]
Incidents Where Perpetrator(s) Were Unknown	76	29	58

SOURCE: POICN Database (Binder and Ackerman, 2020).

TABLE 2-4 Most Prolific Formal Terrorist Organization Perpetrators (All Chemical Incidents)

Group Name	Number of Incidents
ISIS	43
Aum Shinrikyo	28
Chechen Rebels	28
East Turkistan Liberation Organization (ETLO)	21
Khmer Rouge	18
Taliban	18
al-Qa'ida	16
Liberation Tigers of Tamil Eelam (LTTE)	11
Hamas	9
Revolutionary Armed Forces of Colombia (FARC)	9
Kurdistan Worker's Party (PKK)	6
Animal Liberation Front (ALF)	5

SOURCE: POICN Database (Binder and Ackerman, 2020).

TABLE 2-5 Most Prolific Formal Terrorist Organization Perpetrators: Heightened Interest Chemical Incidents Only

Group Name	Number of Incidents
ISIS	38
Aum Shinrikyo	21
Chechen Rebels	16
Taliban	16
al-Qa'ida	13
East Turkistan Liberation Organization (ETLO)	5
Revolutionary Armed Forces of Colombia (FARC)	5
Liberation Tigers of Tamil Eelam (LTTE)	4
al-Qa'ida Organization in the Islamic Maghreb	3
Khmer Rouge	3
National Liberation Army (Colombia) (ELN)	3

SOURCE: POICN Database (Binder and Ackerman, 2020).

TABLE 2-6 Most Prolific Formal Terrorist Organization Perpetrators: Uses

Group Name	Number of Incidents
ISIS	31
East Turkistan Liberation Organization (ETLO)	21
Taliban	17
Aum Shinrikyo	16
Khmer Rouge	15
Liberation Tigers of Tamil Eelam (LTTE)	8
Chechen Rebels	7
Revolutionary Armed Forces of Colombia (FARC)	6
Animal Liberation Front (ALF)	2

SOURCE: POICN Database (Binder and Ackerman, 2020).

Many of the same incentives or disincentives for selecting chemical weapons may apply to nonideological violent nonstate actors (e.g., not strictly terrorists) who are not reflected in the above data. However, external belief systems that can either prompt or constrain weapon selection do not exist for these groups. The goals of nonideological violent nonstate actors are likely to be more personal (Ackerman et. al., 2017a), such as holding a grudge against a particular individual, seeking financial gain, or having psychotic delusions directing them to cause harm. In many of these idiosyncratic cases, understanding and detecting these behaviors, including weapon selection, will be more challenging. It is interesting to note that, unlike ideologically driven nonstate actors, when both terrorist and nonterrorist perpetrators are taken into account ~ 55 percent were lone actors (Ackerman and Binder, 2017b).

2.1.2 Demographic of Perpetrator and Capability

While motivation can be a powerful driving force that can in turn spur the development of a chemical weapon capability, it does not guarantee the acquisition of a viable chemical weapon. Figure 2-5 below depicts the filtering process between an intent to utilize CW and the actual capability to do so, including for heightened interest events. Protoplots (see footnote 2) are excluded, as they do not represent a clear intent to acquire a CW. It is salient to note that the majority of perpetrators, whether working singly or as part of a team, were able to acquire a chemical agent and create a weapon of some sort.

It is essential to understand the spectrum of means that may be employed by terrorists to deliver the chemical agent where it would result in maximum harm so that it can be incorporated into the strategy for prevention and deterrence.

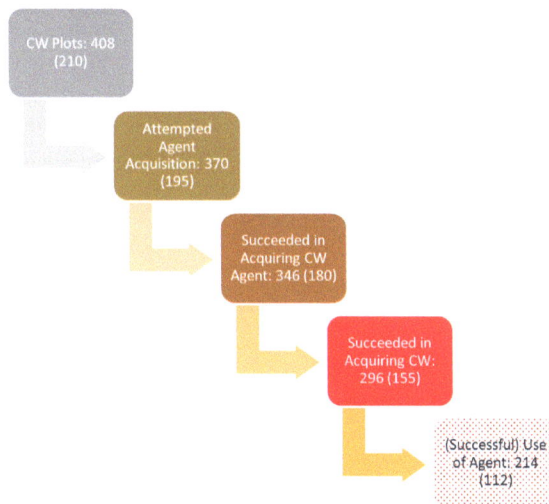

FIGURE 2-5 Flow chart illustrating pathway from chemical weapon plot to acquisition of chemical weapon. Heightened Interest events in parentheses.
SOURCE: POICN Database (Binder and Ackerman, 2020).

A large-scale delivery of a chemical agent, in terms of quantity, is a much greater challenge for a terrorist organization in comparison to the delivery of a smaller quantity. Although it is easier to execute, the smaller-scale delivery may have a smaller impact in terms of creating harm or yielding limited results. Factors such as weather patterns, chemical stability and volatility, and unintentional poisonings influence the success of the delivery.

Aum Shinrikyo (see Appendix E for case study) provides an illustration of the difficulty that a terrorist organization would encounter in attempting to deliver toxic agents on a large scale. Their primary sarin attack on the Tokyo subway utilized the vapor pressure of sarin and nothing else (Tucker, 2006). The terrorists' perforated polyethylene bags containing sarin in the Tokyo subway, exited the subway and relied on the sarin to volatilize on its own. The dispersal method was crude and was not particularly efficient; nevertheless, the release caused 12 fatalities, 54 victims in serious or critical condition, and more than a thousand victims with mild symptoms. If a more sophisticated dispersal method had been used, a significantly larger number of casualties could have resulted. The Aum sarin attack showed that even a well-funded organization that had acquired significant synthetic ability capable of producing sophisticated nerve agents may not necessarily possess sophisticated dispersal technology.

Furthermore, volatile—or to a lesser degree semi-volatile TICs—are a means for an opportunistic chemical warfare agent (CWA) attack by a terrorist organization. Initial utilization of CWA involved the release of chlorine from pressurized steel cylinders during WWI. The approach had the drawback of relying on wind direction and speed to transport the chlorine to enemy lines. However, a terrorist organization might not be as concerned about wind direction since there is less discrimination about who would be exposed. The disaster at Bhopal is an example of the casualties and damage that could be caused by this type of terrorist attack (see Chapter 5 for further details). The release of methyl isocyanate caused >3,700 deaths and injured perhaps another 20,000.

To examine capability factors a little more closely, we can draw on the Chemical and Biological Weapons NonState Adversary Database (CABNSAD, Ackerman and Binder, 2017b), which focuses on the perpetrators themselves and includes both terrorist and nonterrorist violent nonstate actors.[4] The database contains information on 398 individuals involved with chemical weapons incidents of one type or another, with at least 110 incidents perpetrated by non-terrorist actors beyond the 423 incidents recorded in the POICN Database.

With respect to age, the range is 15–70 years old, with a mean age of ~ 37 years and a median age of 34 years. Figure 2-6 breaks down the 217 perpetrators for whom age data is available.

With respect to the highest level of education reached—this information was only available for 84 perpetrators (see Figure 2-7 and Table 2-7). The known disciplinary background of the perpetrators is also shown.

---

[4] The CABNSAD Database comes in two forms, one in which each perpetrator is analyzed individually (even if they were in the same group or cell), and one in which perpetrators within the same organization are aggregated. For the purposes of this section, the nonaggregate version is utilized, since individual-level demographics are presented.

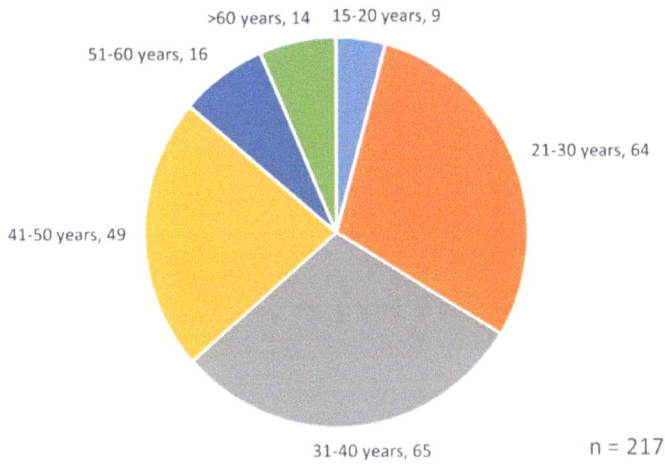

FIGURE 2-6 Number of perpetrators by age.
SOURCE: CABNSAD Database (Ackerman and Binder 2017b).

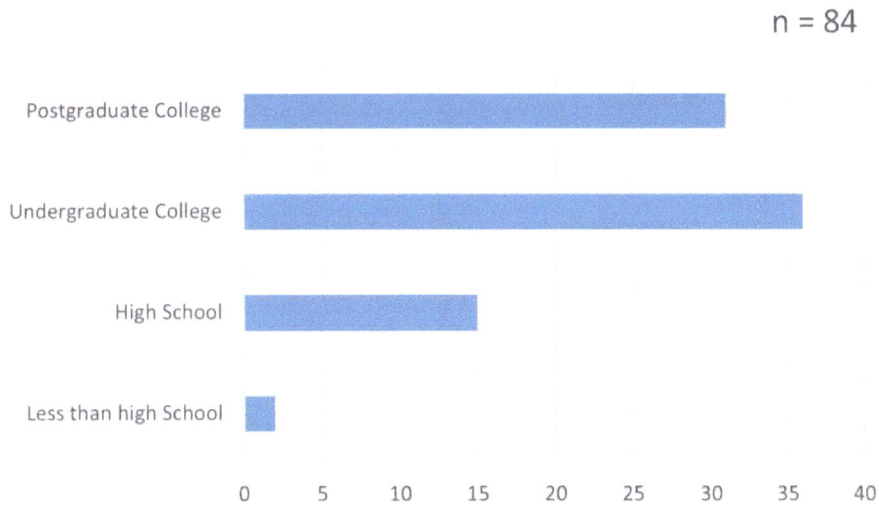

FIGURE 2-7 Perpetrator's highest education level.
SOURCE: CABNSAD Database (Ackerman and Binder 2017b).

TABLE 2-7 Educational Discipline of Perpetrators

Discipline	Frequency
Chemistry and Related Disciplines (incl. Pharmacology)	9
Other Natural Sciences	10
Other STEM (including Engineering)	10
Social Sciences and Humanities	11
Business/Economics	4
Medicine and Related Disciplines	14

NOTE: Where known; n = 58.
SOURCE: CABNSAD Database (Ackerman and Binder, 2017b).

The perpetrator study cited above (Ackerman and Binder, 2017b) derived several quantitative results from the CABNSAD database. While the CABNSAD Database includes perpetrators of both chemical and biological events, the majority of perpetrators in the database pursued chemical weapons. (Only 87 of 486, or ~ 18 percent solely pursued biological weapons.) Additional analyses would need to be conducted to confirm that the following findings apply to chemical perpetrators in particular; it is suspected that many of them will hold for the more limited dataset of chemical perpetrators. The following observations were made based on the combined chemical and biological perpetrator data:

- The majority of individual perpetrators (e.g., not part of a group or cell) reached at least an undergraduate level of education.
- Far more (nonfatal) injuries occurred when the perpetrator had a college education.
- The higher the level of education of the perpetrator, the more likely the perpetrator would successfully use chemical and/or biological agents.
- Successful perpetrators were more likely to be older and involved with chemical and/or biological agents for a longer period of time in comparison to the younger perpetrators.
- Unlike the case with terrorist groups, the majority of perpetrators, overall, targeted food or drink with their chemical and/or biological agents.

2.1.3 Wide-Ranging Baseline Threat

Terrorists employing chemical agents have caused a greater amount of harm than those using any other type of unconventional weapon. They have killed at least 150 people (and possibly over 900) and injured at least 2,400 (up to approximately 6,600) between 1990 and 2020 (POICN Database, Binder and Ackerman, 2020).[5] Nonideologi-

---

[5] There are several attacks, such as one by the Khmer Rouge in 1996 that reportedly killed 200 and wounded 300, which carry some doubt as to their veracity, whereas several other attacks might be viewed more as

cal perpetrators have also killed at least 230 people and injured at least 1,840 over the same period (CABNSAD Ackerman and Binder, 2017b).

Yet, most incidents have resulted in no casualties: ~ 80 percent of chemical perpetrators did not cause any fatalities and ~ 70 percent did not cause any nonfatal injuries. Indeed, the POICN Database could only confirm that attacks with chemical agents were responsible for any fatalities in ~ 12 percent of all terrorist use events and for injuries in only ~ 40 percent of terrorist use events. A handful of perpetrators (fewer than a dozen) have thus been responsible for the majority of casualties and fatalities.

With respect to terrorism, the sarin nerve agent attacks carried out by the Japanese Aum Shinrikyo cult in the mid-1990s (14 dead, 1,050 injured in total, see Gupta, 2015; Smithson and Levy, 2000) have been the most consequential in terms of casualties, and because they occurred almost without warning in the context of a peaceful civil society. More recently, the repeated use by the Islamic State (and its predecessors) of chlorine and mustard gas in Iraq has broadened the ambit of how terrorists might deploy chemical weapons. Aside from possibly the Islamic State (IS), the nonstate actors who have inflicted the greatest amount of lethal harm using chemical agents have been apocalyptic millenarian cults, in particular, the People's Temple of Jim Jones which killed over 900 people (see BBC News) and the Movement for the Restoration of the Ten Commandments which killed over 20 (see Borzello, 2000). However, in these cases, the deployment took the form of poisoning their own members.

Thus, while jihadist groups like IS have recently demonstrated the highest threat potential for chemical terrorism, historically apocalyptic cults have caused the most casualties (usually to their own members). Moreover, there has been a wide range of ideologies and actors that have pursued attacks with chemical agents. Formal organizations may dominate the chemical terrorism landscape, but overall individuals are playing a larger role in this space. Chemical terrorism, at least at a nominal level, appears to be achievable by many violent actors, with over half of these plots having proceeded all the way to the use stage. Although the vast majority of attacks have involved lower-toxicity agents and crude delivery methods; both warfare agents and other high-toxicity chemicals, and sophisticated delivery mechanisms have been pursued and employed.

FINDING 2-1: The current threat landscape consists of multiple lower-consequence attacks, punctuated by the possibility of occasional large-scale, potentially mass-casualty events.

2.2 CHARACTERIZATION OF BROAD CHEMICAL THREATS

Chemical threat agents are highly hazardous or toxic chemicals that can be acquired or developed as weapons of mass destruction to promptly cause casualties.

---

insurgent attacks rather than terrorism proper. We have provided the most conservative estimate above as a lower bound, but these might significantly undercount the true number of casualties. These figures also do not include injuries from several more recent attacks which are still being assessed by POICN coders, so the true figures might be considerably higher.

The widespread availability of starting materials for millions of highly toxic compounds and instructional materials outlining how to produce them have reduced barriers to entry for the nefarious use of chemicals. Furthermore, with increased industrialization, other commercially available chemicals and materials with the potential to be used as weapons could be acquired and accessed by terrorist individuals and organizations to be used as improvised weapons. The list of known and potential chemical threat agents is vast and expanding (Figure 2-8).

To increase the likelihood of success in countering the continually expanding list of potential chemical threat agents, federal agencies are increasingly turning toward broadly extensible strategies and agent-agnostic approaches (further discussion can be found in section 7.6).

Over 100 billion chemicals exist in the theoretical "molecular universe" (Reymond, 2015). Over 200 million chemicals have been synthesized or isolated,[6] and another is identified every 3–4 seconds (CAS, n.d.; Mulvaney, 2017). Technological advances such as synthetic biology and cheminformatics, additive manufacturing, nanotechnology, and microscale chemical reactors further facilitate the discovery of new and novel chemical threat agents[7] available for potential beneficial or nefarious use.

Against this backdrop, one should consider the state of chemical weapons in context. In this regard, steady progress toward the elimination of declared chemical weapons stockpiles has also driven the research, development, and deployment of new capabilities to detect and respond to a broad range of classes of chemicals with known potential to be used as weapons. However, the norms against weaponizing chemicals enshrined within the Chemical Weapons Convention (CWC) are challenged through suspected states not abiding by their treaty commitments or remaining outside the treaty and uses of chemicals as weapons in ways not broadly anticipated at the time of treaty negotiation. In a globally connected world, the varying robustness and effectiveness of regulation is also a challenge. The U. S. Government (USG) faces a challenge to ensure readiness to prevent, counter, and respond to chemical threats, as the number and complexity of such threats are continually evolving and expanding. Meanwhile, regulatory agencies such as the U.S. Environmental Protection Agency (EPA), the Food and Drug Administration (FDA), or the Drug Enforcement Agency (DEA) can take years to review the safety and toxicity profiles of a new chemical within the United States. Existing strategies and associated capabilities and infrastructure must be reexamined and potentially retooled to stay ahead of the threat.

---

[6] Compendium of WHO and Other UN Guidance on Health and Environment. https://www.who.int/tools/compendium-on-health-and-environment/chemicals.

[7] As stated in Chapter 1, chemical terrorism threats considered include agents identified as chemical weapons as well as existing, emerging, and potential agents of concern. Threat actors' patterns of use were considered to identify trends and to understand the degree to which different methods impacted successful implementation of a given strategy.

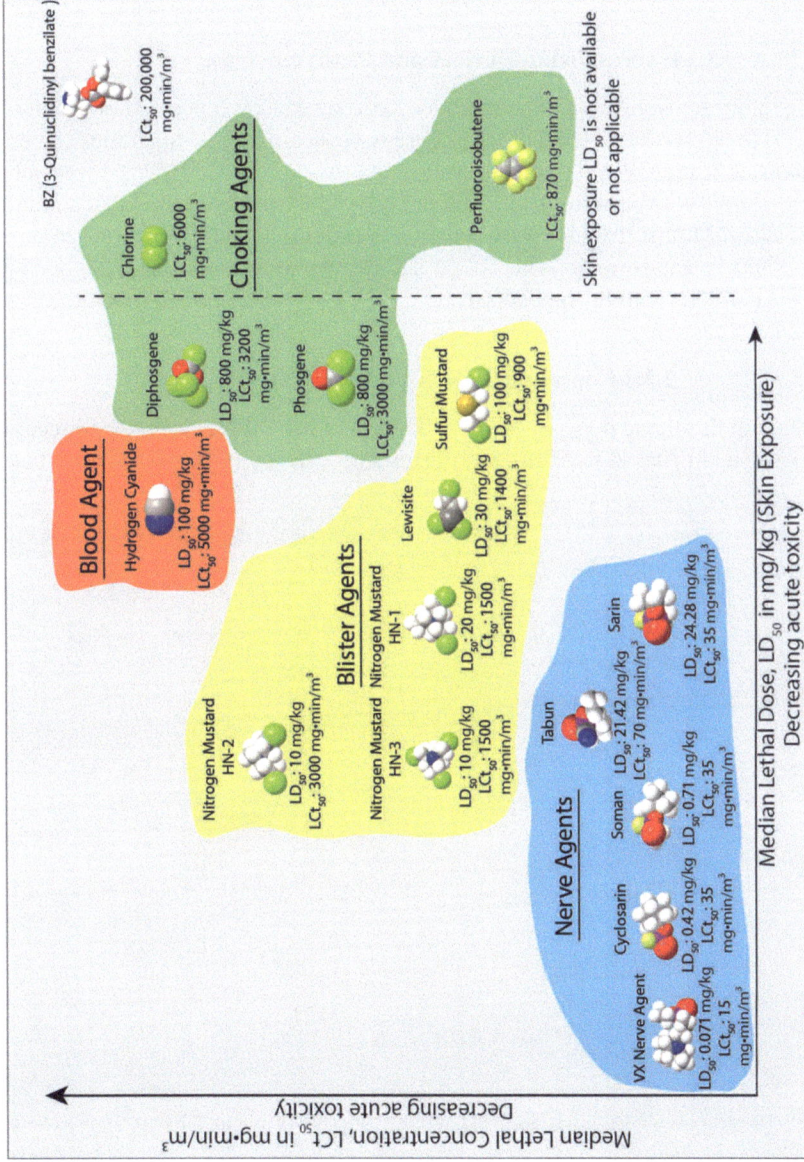

FIGURE 2-8 Chemical warfare agents are broadly categorized according to their effect: the less toxic a chemical, the higher the lethal dose or concentration.
SOURCE: Fischer et al., 2017.

2.3 DELIVERY METHODS OF CHEMICAL AGENTS

Between the period of 1990 and 2020, the incidents of chemical terrorism have included a variety of delivery methods, such as explosive devices, aerosol, and other methods (see Figure 2-9). This section describes the delivery methods that most concern chemical terrorism: passive release, aerosolizing devices, and contamination of food or water.

2.3.1 Aerosolizing Devices and Passive Release

If in the future aerosolization of acute stable toxic substances (chemical or biological) is deemed to be possible, then there is a cause for concern if large quantities can be rapidly dispersed. If the CWA has ideal physical and toxic properties (e.g., surfactant, acutely toxic, stability) to be aerosolized, and can be spread aerially in a manner similar to the application of forest fire suppressing foams over dense urban population centers, the impact could be catastrophic. The Tokyo subway release falls in the category of passive (see Aum Shinrikyo in Appendix E).

2.3.2 Contamination of Food or Water

The nature of this threat depends on the properties of the CWA to cause an immediate and widespread impact. A previous analysis of an accidental poisoning of livestock

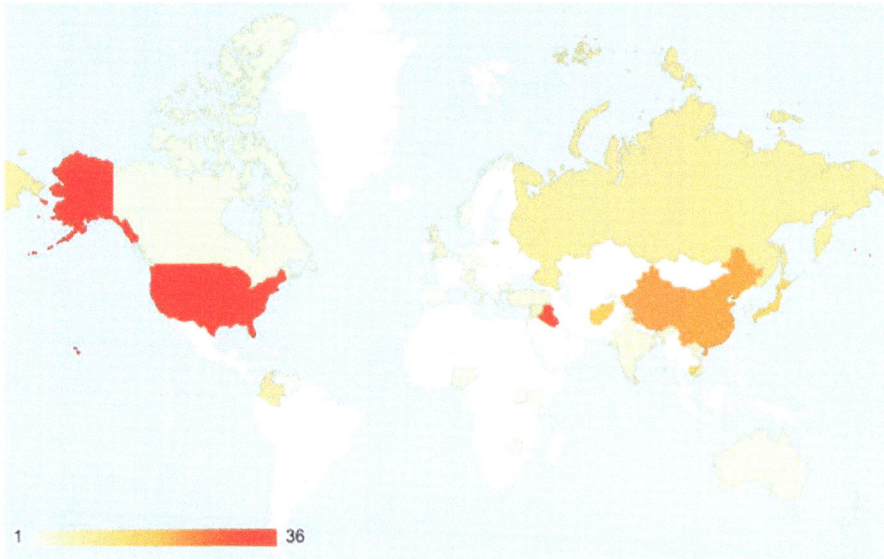

FIGURE 2-9 Geographic distribution showing target countries of chemical terrorism attacks recorded from 1990 to 2020.
SOURCE: POICN Database (Binder and Ackerman, 2020).

feed considered the food safety and security implications in the context of chemical terrorism (Kosal and Anderson, 2004). An example of a historical incident of intentional contamination of food comes from post-WWII Germany. In 1946, a group of Jewish Holocaust survivors poisoned the bread of Nazi S.S. officers in an American Prisoner of War (POW) camp (Tucker, 2006; The Guardian, 2016). An arsenic-containing material was used to poison black rye bread that was to be served to the detainees. More recent domestic cases include contamination of ground beef with a nicotine-based pesticide in a Southwest Michigan supermarket in December 2002 (Dasenbrock et al., 2005) and intentional contamination of coffee with arsenic after a church service in Maine (Dasenbrock et al., 2005).

Contamination of water, especially large municipal supplies (e.g., reservoirs) is beyond the capabilities of most terrorist groups due to dilution factors. The potential challenges at the point of distribution were illustrated by the widely reported cyber hacking incident of a water treatment facility in Florida. After investigating, the FBI "was not able to confirm that this incident was initiated by a targeted cyber intrusion" (Vasquez, 2023). It appears more likely that it was an employee error (Cohen, 2021; Teal, 2023).

2.4 EMERGING CHEMICAL THREAT TECHNOLOGIES

2.4.1 Artificial Intelligence/Machine Learning (AI/ML) and Quantum Computing

Computational chemistry is a combination of chemometrics, cheminformatics, and modeling. Chemical properties and features that affect the physiological activity of a compound can be predicted by combining chemometrics, cheminformatics, and quantum-structure activity relationship (QSAR)-based ML models (Figure 2-10).

Against this backdrop, a number of researchers have demonstrated the use of such capabilities to predict novel VX-related nerve agents (Urbina et al., 2022), vapor pressure for unascertained Novichoks (Jeong et al., 2022), and novel protein structures (Anishchenko et al., 2021). However, while such tools significantly lower barriers to entry for a motivated actor to design highly toxic compounds in silico, their use to actually enable a terror attack involving chemicals would still also require sophisticated chemistry, engineering expertise, materials to synthesize the candidate compounds, and to handle them safely until deployed—a difficult task for known chemicals, but which becomes more difficult for novel chemicals whose properties have not yet been studied (Nasser et al., 2022; Urbina et al. 2022). A number of factors combine to minimize the likelihood of terrorist use of such tools, such as (1) low incentive because other highly toxic chemicals that can be used in chemical attacks are more readily available, (2) limited access to specialized AI/ML capabilities, and (3) limited access to chemical precursors needed for synthesis. Additionally, large language models have modest safeguards against creating recipes for chemical weapons, and if those are circumvented, the poor quality of information on the web makes it unclear that these models will increase the risks of effective chemical terrorism. It is imperative that assessments on emerging technologies and their roles in chemical terrorism are grounded in both technical and operational rigor.

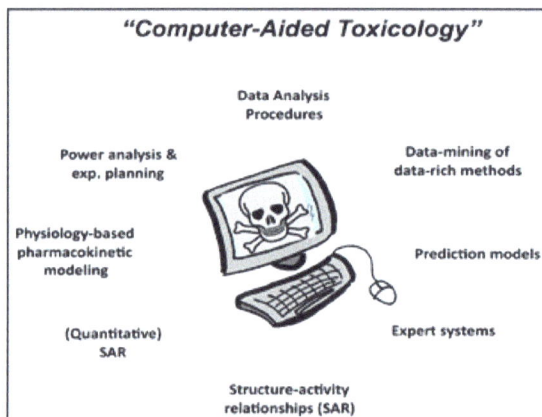

FIGURE 2-10 Advances in ML, explainable AI, chemometrics, and a variety of computational chemistry methods allow for efficient computer-aided toxicological prediction and design of thousands of candidate chemical compounds with specific physiological activities.
SOURCE: Hartung and Hoffman, 2009.

Advances in AI/ML are also being pursued to attempt to model and predict terrorist behavior and tactics (Uddin et al., 2020). Many of these attempts leverage social media (Al-Shaibani and Al-Augby, 2022), satellite imagery (Buffa et al., 2022), and other larger data sets (Krieg et al., 2022).

CONCLUSION 2-2: While AI and ML can be used to predict new chemical structures, the feasibility of converting predicted structures to weaponized chemicals is not straightforward and is thus unlikely in the near term (~ 5 years).

2.4.2 Synthetic Biology and the Chem-Bio Interface

For the potential applications of synthetic biology (i.e. "SynBio") to chemical terrorism threats, three important concepts must be acknowledged (Kosal, 2021). First, SynBio is not a discrete homogenous thing. One of the most well-known tools of SynBio is the advanced gene editing technique known as CRISPR, which is a bacteria-derived system that uses specific proteins, such as Cas9, to "cut and paste" selectively into a genome. Applications of this technique include, for example, crop pesticide resistance, disease treatments, and biofuels. SynBio is a concern for chemical terrorism because if a toxic genetic material is coupled with an efficient delivery method then the outcome could potentially be lethal, although currently SynBio's implications on security and safety remain largely uncertain.

First, Cas9 is not the only protein (e.g., Cas12 and Cas13), and CRISPR is not the only advanced gene editing system available. There is no single SynBio system to target when trying to assess potential threats. Even the "easiest" CRISPR synthesis is harder and requires more tacit knowledge and specialized equipment than the construction of an improvised explosive device (IED). The knowledge and skill required are well within the capacity of most states and large transnational corporations; it is unlikely in the case of terrorists (Kosal, 2021).

Another example of a SynBio technique or process (that gets less popular attention) is cell-free synthesis (CFS) (Vilkhovoy et al., 2020). It is a method or platform to produce something, mostly small molecules like chemicals and proteins. CFS can serve as a replacement platform or alternative production means for something when cell-based systems are problematic (Lu, 2017; Vilkhovoy et al, 2020). Metabolic engineering and microbial cell factories (MCF) are another aspect of SynBio that bridges technique, process, and products. Microbial cell factories are means to produce materials, such as on-site synthesis of fuels, specialty chemicals, or other commodities that are not dependent on petrochemicals (Amer et al., 2020; Jiang et al., 2020; Linger et al., 2020; Nawab et al., 2020; Yan and Pfleger, 2020). **In the committee's judgment, terrorists are unlikely to pursue CFS or MCFs as a means to synthesize chemicals due to the sensitive and controlled conditions (temperature, media, enzymes) that are required for high-yielding growth.**

Second, breakthroughs and discoveries in SynBio come from molecular biology, chemistry, physics, and multiple engineering fields; thus there is no single scientific discipline on which to focus security attention when considering the implications of this area.

Finally, and perhaps adding the greatest amount of complexity, synthetic biology is fundamentally "dual-use" in its nature. It is a dual-use technology, by both meanings of the term. Historically and in the nuclear policy world, dual-use means a demarcation between civilian and military uses. In the life sciences and much of cutting-edge scientific and engineering research, dual-use refers to the concept that the same or similar techniques, manufacturing elements, and processes used for beneficial purposes could also be misused for deleterious purposes. Almost all the equipment and materials needed to develop dangerous or offensive agents, particularly biological and chemical agents, have legitimate uses in a wide range of scientific research and industrial activity, including defensive military uses. Advances in synthetic biology and gene editing not only potentially pose security and proliferation concerns, but they also may enable new capabilities for defense, detection, and verification of chemical and biological agents. These advancements are, in addition to their important role in enhancing diagnostic capabilities for emerging infectious diseases, like COVID-19, have multiple other beneficial outcomes beyond therapeutic gene editing (Kosal, 2021). The dual-use nature of SynBio—and much of modern science and technology (S&T)—adds further complications to identification, countering, and response.

2.4.3 Advanced Materials Science

Advances in materials science have enabled the production of tailored materials where specific particle size, surface chemistry, porosity, and other properties can be purposefully produced. The interaction of particles with chemicals can repress (e.g., permeation barrier/timed release) or enhance (e.g., aerosolization aid) dispersal. Advanced materials for drug delivery are designed to move therapeutics across barriers such as the mucosal barrier (Bandi et. al., 2021; Dong, 2018; Kosal, 2009). The same advanced materials and their underlying mechanisms that can be used to deliver beneficial therapeutics could be used for nefarious purposes to deliver toxic entities. The use of advanced materials represents a significant technological, financial, and personnel investment and thus has many barriers to deployment.

CONCLUSION 2-3: Advanced material science could enhance an attack but is much more of a concern with respect to significant large and well-resourced state-based programs than for terrorism in the near to medium term.

2.4.4 Small-Scale Reactors

Microfluidics or microreactors or "labs on a chip" involve chemical reactions taking place in very small, enclosed spaces. Picture a computer chip, but instead of wires, it has tiny channels, roughly the size of a human hair, where liquids flow and mix, though devices can also take other forms. Similarly, nanofluidics, or nanoreactors, involve channels an order of magnitude smaller, which potentially enables very precise control of reactions and their thermal output.

These advances have implications for both the chemical weapons threat and responses to it. For example, on the offensive side, it might enable the rapid synthesis and testing of novel chemical weapons agents or covert, on-demand production of threat agents. Defensively, it is already enabling more effective sensing devices and so-called organ-on-a-chip devices that facilitate research and development of new medical therapeutics.

In very different ways, advances in microfluidics may have implications for biological terrorism and, in more limited ways, nuclear terrorism, both of which are beyond the scope of this study. The term "microreactor" is also used to describe small nuclear power reactors (DOE, 2021), which have nothing to do with micron-level chemistry or biology.

Today, utilizing microfluidic devices requires more sophistication than utilizing traditional laboratory processes, and thus these techniques are mostly the domain of states or advanced corporations. In the future, as with many other advanced technologies, the dynamic is likely to invert, and the technology will likely become more turnkey, i.e., could be used with little or no understanding of what processes were actually occurring inside a device (just as someone needs no understanding of how the computer chips and other components in a smartphone operate to use one). Of course, even if in the future the technology supports such turnkey applications, this does not mean the market will provide them, or states will be powerless to regulate them. One application that has had limited investigation in the context of possible diversion from humanitarian or international development programs for misuse is the specialized field of frugal science, which often incorporates microfluidic elements (Tennenbaum and Kosal, 2021).

These advances constitute a powerful set of tools, but they also have limitations. One major limitation is corrosion, which is a challenge due to extensive surface area contact, often high chemical throughput, and delicate structures, mediated both by the chemicals being used and the materials out of which the structures are composed. Another challenge is that precipitation reactions are unsuitable for microchemistry since solid precipitates can clog small channels (Zong and Yue, 2022). Among present-day chemical agents, microfluidics could be employed to produce some agents but not others.

In the chemical terrorism context, today microfluidic devices may bolster state efforts to respond to the threat, especially at the basic and applied research level. One could imagine extremely sophisticated nonstate actors, perhaps with relevant government or private sector experience, employing the technology offensively, but that seems relatively unlikely in the near future. The one potential caveat here is that state-supported terrorists might benefit from offensive applications of the technology. Down the road, it is at least conceivable that such advances might be employed by terrorists without state support, depending on how the technology and its accessibility evolve.

> **RECOMMENDATION 2-4: The intelligence community (IC) should continue to monitor interest in emerging technologies and delivery systems, such as drug delivery systems, and trends by terrorist groups to innovate and improvise. This is likely to look significantly different than the applications of advanced materials chemistry by GPS.**

2.5 EMERGING ACTORS

Some terrorists, especially those of the mid-twentieth century, have used violence to seek political change and wanted to take control of political processes through violence, often in the context of anti-colonial or separatist efforts (Rapoport, 2004). Others, such as many early twenty-first-century terrorist groups, "don't want a seat at the table, they want to destroy the table and everyone sitting at it."[8] In either case, the political element of terrorism remains. The nature of terrorism, evolution and dynamics of terrorist groups, geopolitics of responses, and broader study of topics related to terrorism are vibrant areas of scholarly work (Schuurman, 2020). In assessing the adequacy of current strategies to counter chemical terrorism, a brief overview of major trends of would-be perpetrators of chemical terrorism and some of the thinking on the future of chemical terrorism is warranted.

For the foreseeable future, the character of chemical terrorism-related threat actors seems likely to remain relatively constant or to evolve slowly. Yet, sharp discontinuities cannot be ruled out and are difficult to predict. If threat actors evolve significantly, this is likely to occur because of perceived changes in the perceived tactical and/or strategic benefits of using chemical agents, changes in ideology, or changes in various idiosyncratic factors related to motivation. Changes in the perceived benefits of chemical agents might also be related to the perceived *lack* of efficacy of other attack modes; in other words, chemical weapons might be so-called "weapons of the weak" to which actors who lack other alternatives turn. Certain types of terrorist groups—highly religiously identifying, apocalyptic, and right-wing, anti-government groups—have been shown to

---

[8] Former CIA Director James Woolsey quoted in https://www.newyorker.com/magazine/2001/10/29/what-terrorists-want.

have a greater propensity to use chemical (and biological) agents (section 2.1; Tucker, 2000). Understanding which types of groups and motives might be associated with future pursuit or use of chemical agents is important.

Potential changes in technology and their effects on the capabilities required to conduct chemical terrorism threats are dealt with above in section 2.4. However, it is worth noting that changes in technology might also spur changes in motivation. For example, if technological change makes certain forms of chemical terrorism more feasible, or, conversely, if technological change makes certain forms of nonchemical terrorism less feasible, that might lead previously less motivated actors to revisit their stances.

To date, there has often been an inverse relationship between various nonstate actors/terrorists' motivations and capabilities to conduct CBRN terrorism, including chemical terrorism. Actors who are more capable are often less motivated to pursue CBRN terrorism. One reason for this is that more capable terrorists often fall into this category because they benefit from state sponsorship. They might thus be constrained by their state patrons. Conversely, actors who are more motivated to pursue unconventional weapons like chemical agents, are often less capable. One potential source of discontinuity in the future is a change in this dynamic, for example, more capable actors developing greater motivation or a more motivated actor developing greater capability.

Another potential source of discontinuity lies in the nexus between terrorism and crime. Terrorist groups often engage in crime to support their activities.[9] Criminal groups sometimes morph into terrorist groups.[10] And, of course, terrorists and criminals sometimes collaborate.[11] The potential effects of chemical terrorism risk are various

---

[9] There are many examples. On the Liberation Tigers of Tamil Eelam (LTTE), see Hutchinson, Steven, and Pat O'Malley. 2007. "A Crime–Terror Nexus? Thinking on Some of the Links between Terrorism and Criminality." *Studies in Conflict & Terrorism* 30(12): 1095–1107. doi: 10.1080/10576100701670870). LTTE happens to be the first confirmed case of nonstate actors using chemical weapons in warfare; see "The First NonState Use of a Chemical Weapon in Warfare: The Tamil Tigers' Assault on East Kiran." *Small Wars & Insurgencies:* (20)3–4. Another example is Al Shabaab; see Petrich, Katharine. 2022. "Cows, Charcoal, and Cocaine: Al-Shabaab's Criminal Activities in the Horn of Africa." *Studies in Conflict & Terrorism*, 45(5–6): 479–500. doi: 10.1080/1057610X.2019.1678873).

[10] On criminals morphing into terrorist groups, examples include D Company; see Clarke, Ryan and Stuart Lee.2008. "The PIRA, D-Company, and the Crime-Terror Nexus." *Terrorism and Political Violence* (20)3: 376–395. doi: 10.1080/09546550802073334. al-Qa'ida in the Lands of the Islamic Maghreb's (AQIM) offshoot Jama'at Nusrat al-Islam wal-Muslimin (JNIM) has recruited criminals into its organization; see Beevor, Eleanor. 2022. "JNIM IN BURKINA FASO: A Strategic Criminal Actor." *Global Initiative Against Transnational Organized Crime*. https://globalinitiative.net/wp-content/uploads/2022/08/Burkina-Faso-JNIM-29-Aug-web.pdf). Mexican drug trafficking organizations have engaged in what is commonly labeled narcoterrorism; see Phillips, Brian J. 2018. "Terrorist Tactics by Criminal Organizations: The Mexican Case in Context." *Perspectives on Terrorism* 12(1): 46–63. http://www.jstor.org/stable/26343745.

[11] On cooperation between criminal organizations and terrorist groups, examples include Ndrangheta and al-Qa'ida; see Makarenko, Tamara and Michael Mesquita. 2014. "Categorizing the Crime–Terror Nexus in the European Union." *Global Crime*. 15(3-4): 259 –274. doi: 10.1080/17440572.2014.931227. Other examples include Basque Euskadi Ta Askatasuna (ETA) and Italian criminal organizations or the Revolutionary Armed Forces of Colombia–People's Army (FARC) and Colombian drug cartels; on both, see Hutchinson, Steven, and Pat O'Malley. 2007. "A Crime–Terror Nexus? Thinking on Some of the Links between Terrorism and Criminality." *Studies in Conflict & Terrorism* 30(12): 1095–1107. doi: 10.1080/10576100701670870.

and nuanced, given that transnational criminal groups are heavily involved in the drug trade, including the smuggling and production of synthetic narcotics.

An important dynamic, which could have either modest or more significant impacts, is demonstration effects. Terrorism tactics sometimes exhibit copycat dynamics or go "viral," the classic example being suicide bombing. A chemical terrorist attack that is perceived as successful might motivate others to try to do likewise (see Dugan, LaFree, and Piquero, 2005; Guelke, 2017; Horowitz, 2010). Of course, the opposite is possible, too, i.e., observing a failed or ineffective chemical terrorism attack (and government or public responses to it) might motivate others to *not* pursue chemical terrorism.

The Islamic State's recent use of crude chemical agents, to little effect, seems unlikely to motivate others to try to do likewise, nor are most other extant terrorist actors likely to achieve even the limited successes IS did. It is also striking that IS mostly used chemical agents in combat operations, rather than against civilian targets. Further, while IS publicly celebrated various forms of hyperviolence—like the burning to death of a captured Jordanian pilot—it continued to deny its own possession and use of chemical weapons.

On the other hand, the Syrian, Russian, and North Korean uses of chemical agents in recent years, in both indiscriminate and targeted ways, could perhaps inspire others (including nonstate actors) to try to do likewise. Indeed, another potential source of chemical terrorism-related discontinuities is state behavior. Whether any state behavior, even directly attacking civilians, is appropriately labeled terrorism is contested, though some would make the case that the label applies. And even if we limit terrorism to violence perpetrated by nonstate actors, states can be sources of both witting and unwitting aid to nonstate actors. With respect to the future threat of chemical terrorism in the near term:

- The most significant international chemical terrorism threats are likely to continue to be from jihadists, while the most significant domestic chemical terrorism threats facing the United States are likely to continue to be from far-right extremists.
- Cults are a potential source of threats, too, and given their insular nature and potential for rapid and extreme radicalization, the threat they pose is both difficult to predict and difficult to detect if it manifests.
- Lone actors appear likely to continue to be responsible for a large number of incidents but only modest consequences, with motivations including political/ideological, criminal, and mental health related.
- Whether and how certain emerging ideological milieus will interact with chemical terrorism remains unclear. This includes incels (men who identify as involuntary celibates and blame both women and certain other men), accelerationist[12] dynamics that cut across various right-wing extremist milieus (with bolder goals and an associated willingness to take bolder action in service

---

[12] See https://www.middlebury.edu/institute/academics/centers-initiatives/ctec/news/ctec-expert-philipp-bleek-recently-presented-threat-far.

of them), and anti-technology/industrialization extremists (including, but not limited to, certain environmental extremists).[13]

REFERENCES

Ackerman, G., C. Davenport, M. Binder, H. Tinsley, R. Earnhardt, C. Watson, M. Watson, and T. K. Sell. 2017a. *Profiling the CB Adversary: Motivation, Psychology, and Decision*. College Park, MD: START.

Ackerman, G., and M. Binder. 2017b. *Chemical and Biological Non-State Adversaries Database (CABNSAD)*. College Park, MD: START.

Al-Shaibani, H. A. A., and S. Al-Augby. 2022. "Terrorist Tweets Detection using Sentiment Analysis: Techniques and Approaches." In *2022 5th International Conference on Engineering Technology and its Applications (IICETA)*, 585–590. Al-Najaf, Iraq. https://doi.org/10.1109/IICETA54559.2022.9888461.

Amer, M., H. Toogood, and N. S. Scrutton. 2020. "Engineering Nature for Gaseous Hydrocarbon Production." *Microbial Cell Factories* 19(1): 209.

Anishchenko, I., S. J. Pellock, and T. M. Chidyausiku. 2021. "De Novo Protein Design by Deep Network Hallucination." *Nature* 600(7889): 547–552. https://doi.org/10.1038/s41586-021-04184-w.

Bale, J. M. 2017. *The Darkest Sides of Politics, II: State Terrorism, "Weapons of Mass Destruction," Religious Extremism, and Organized Crime*. Routledge.

Bale, J. M., and Ackerman G. A. 2009. "Profiling the WMD Terrorist Threat." In *WMD Terrorism: Science and Policy Choices*, edited by Stephen M. Maurer, 11–46. Cambridge, MA: MIT Press.

Bandi, S. P., S. Bhatnagar, and V. V. K.,Venuganti. 2021. "Advanced Materials for Drug Delivery across Mucosal Barriers." *Acta Biomaterialia*, 119: 13–29. https://doi.org/10.1016/j.actbio.2020.10.031.

BBC News. 1978. "Mass Suicide Leaves 900 Dead."

Binder, M., and G. Ackerman. 2020. Profiles of Incidents Involving CBRN and Non-State Actors (POICN) Database. National Consortium for the Study of Terrorism and Responses to Terrorism. College Park, MD.

Binder, M., Gary Ackerman, Cory Davenport, Herbert Tinsley, Rebecca Earnhardt, Crystal Watson, Matt Watson, and Tara Kirk Sell. 2017. "Profiling the CB Adversary: Motivation, Psychology and Decision." START College Park, MD. September.

Binder, M. K. and G. A. Ackerman. 2019. "Pick Your POICN: Introducing the Profiles of Incidents involving CBRN and Non-State Actors (POICN) Database." *Studies in Conflict & Terrorism* (March). https://www.tandfonline.com/doi/full/10.1080/1057610X.2019.15/77541.

Borzello, Anna. 2000. "Mass Graves Found in Sect House." *Guardian*, March 25, 2000.

Buffa, C. V. Sagan, G. Brunner, and Z. Phillips. 2022. "Predicting Terrorism in Europe with Remote Sensing, Spatial Statistics, and Machine Learning." *ISPRS International Journal of Geo-Information* 11(4): 211. https://doi.org/10.3390/ijgi11040211.

Cameron, G. 1999. *Nuclear Terrorism: A Threat Assessment for the twenty-first century*. New York: Macmillan Palgrave.

CAS. (n.d.). CAS Registry. https://www.cas.org/cas-data/cas-registry.

---

[13] See https://www.tandfonline.com/doi/full/10.1080/1057610X.2018.1471972.

Casillas RP, Tewari-Singh N, Gray JP. "Special Issue: Emerging Chemical Terrorism Threats." *Toxicol Mech Methods*. 2021 May;31(4):239-241. doi: 10.1080/15376516.2021.1904472. PMID: 33730980; PMCID: PMC10728888.

Caves, J. Jr., and W. S. Carus. 2021. "The Future of Weapons of Mass Destruction. An Update." A National Intelligence University Presidential Scholar's Paper. National Intelligence University. February.

Cohen, G. 2021. "Throwback Attack: An Insider Releases 265,000 Gallons of Sewage on the Maroochy Shire." *Industrial Cybersecurity Pulse*. https://www.industrialcybersecuritypulse.com/facilities/throwback-attack-an-insider-releases-265000-gallons-of-sewage-on-the-maroochy-shire.

CRS (Congressional Research Service). 2006. CRS Report for Congress, Chemical Facility Security. August 2, 2006.

Dasenbrock, C.O., L. A. Ciolino, C. L. Hatfield, and D. S. Jackson. 2005. "The Determination of Nicotine and Sulfate in Supermarket Ground Beef Adulterated with Black Leaf 40." *Journal of Forensic Sciences* 50(5) (September): 1134–1140. PMID:16225221.

DoD/CBDP (Department of Defense/Chemical and Biological Defense Program). 2023. Approach for Research, Development, and Acquisition of Medical Countermeasure and Test Products. https://media.defense.gov/2023/Jan/10/2003142624/-1/-1/0/approach-rda-mcm-test-products.pdf.

DOE (U.S. Department of Energy). 2021. What is a Nuclear Microreactor? Office of Nuclear Energy. February 26, 2021. https://www.energy.gov/ne/articles/what-nuclear-microreactor.

DOJ (U.S. Department of Justice). 2010. Amerithrax Investigative Summary. February 19, 2010, https://www.justice.gov/archive/amerithrax/docs/amx-investigative-summary.pdf; Bunn and Sagan (eds) *Insider Threats* (book), chapter on Amerithrax.

Dolnik, A. 2007. *Understanding Terrorist Innovation: Technology, Tactics and Global Trends*. New York: Routledge.

Dong, X. 2018. "Current Strategies for Brain Drug Delivery." *Theranostics* 8(6): 1481–1493. https://doi.org/10.7150/thno.21254.

Fischer, E., M.-M. Blum, W. S. Alwan, and J. E. Forman. 2017. "Sampling and Analysis of Organophosphorus Nerve Agents: Analytical Chemistry in International Chemical Disarmament." *Pure and Applied Chemistry* 89(2): 249–258. https://doi.org/10.1515/pac-2016-0902.

The Guardian. 2016. "Failed Jewish Holocaust Survivor Plot to Kill Nazis Still a Mystery after 70 Years." www.theguardian.com/world/2016/aug/31/jewish-holocaust-survivor-kill-nazis-poison-arsenic-nuremburg.

GAO (United States Government Accounting Office). 2020. Report to Congressional Requestors, Chemical Security. DHS Could Use Available Data to Better Plan Outreach to Facilities Excluded from Anti-Terrorism Standards. September 2020.

Gupta, R. C. 2015. *Handbook of Toxicology of Chemical Warfare Agents*. Academic Press. ISBN 978-0-12-800494-4.

Hartung, T., and S. Hoffmann. 2009. "Food for Thought on in Silico Methods in Toxicology." *ALTEX —Alternatives to Animal Experimentation* 26(3):155–166. https://doi.org/10.14573/altex.2009.3.155.

Hoffman, B. 1992. *Terrorist Targeting: Tactics, Trends, and Potentialities*. Santa Monica, CA: RAND.

Jenkins, B. 1986. "Defense Against Terrorism." *Political Science Quarterly 101, Reflections on Providing for "The Common Good"* 101(5): 777–778.

Jeong, K., J. Y. Lee, S. Woo, D. Kim, Y. Jeon, T. I. Ryu, S. R. Hwang, and W. H. Jeong. 2022. "Vapor Pressure and Toxicity Prediction for Novichok Agent Candidates Using Machine Learning Model: Preparation for Unascertained Nerve Agents after Chemical Weapons Convention Schedule 1 Update." *Chemical Research in Toxicology* 35(5): 774–781. https://doi.org/10.1021/acs.chemrestox.1c00410.

Jiang, T., C. Li, Y. Teng, R. Zhang, and Y. Yan. 2020. "Recent Advances in Improving Metabolic Robustness of Microbial Cell Factories." *Current Opinion in Biotechnology* 66: 69–77.

Kosal, M. E. 2006. "Terrorism Targeting Industrial Chemical Facilities: Strategic Motivations and the Implications for U.S. Security." *Studies in Conflict and Terrorism* 30(1): 41–73.

Kosal, M. E. 2009. *Nanotechnology for Chemical and Biological Defense*. New York: Springer Academic Publishers. http://www.springer.com/us/book/9781441900616.

Kosal, M. E. 2021. "CRISPR & New Genetic Engineering Techniques: Emerging Challenges to Nonproliferation.," *Nonproliferation Review*. 27(4–6), 389–408. https://www.tandfonline.com/doi/full/10.1080/10736700.2020.1879464.

Kosal, M. E. 2021. "Assessing Threats of SynBio–Three Challenges." *NCT Magazine*, July 2021. https://nct-magazine.com/nct-magazine-july/assessing-threats-of-synbio-three-challenges.

Kosal, M. E, and D. E. Anderson. 2004. "An Unaddressed Issue of Agricultural Terrorism: A Case Study on Feed Security." *Journal of Animal Science* 82(11): 3394–3400. https://doi.org/10.2527/2004.82113394.

Krieg, S. J., C. W. Smith, R. Chatterjee, and N.V. Chawla. 2022. "Predicting Terrorist Attacks in the United States Using Localized News Data." *PLoS ONE* 17(6): e0270681. https://doi.org/10.1371/journal.pone.0270681.

Lifton, R. J. 2007. "Destroying the World to Save It." In *Voices of Trauma*, edited by B. Drożdek and J. P. Wilson. Boston, MA: Springer. https://doi.org/10.1007/978-0-387-69797-0_3.

Linger, J. G., L. R. Ford, K. Ramnath, and M. T. Guarnieri. 2020. "Development of Clostridium Tyrobutyricum as a Microbial Cell Factory for the Production of Fuel and Chemical Intermediates From Lignocellulosic Feedstocks." *Frontiers in Energy Research* 8: 183. https://doi.org/10.3389/fenrg.2020.00183.

Lu, Y. 2017. "Cell-Free Synthetic Biology: Engineering in an Open World." *Synthetic and Systems Biotechnology* 2(1) (March): 23–27.

Mulvaney, K. "Chemical Pollution is Soaring Faster than we can Measure it." Seeker. Published February 1, 2017. Accessed October 20, 2021. https://www.seeker.com/chemical-pollution-is-soaring-faster-than-we-can-measure-it-2231114982.htm

Nasser, L. 2022. "40,000 Recipes for Murder." Radiolab. https://radiolab.org/episodes/40000-recipes-murder.

Nawab, S., N. Wang, X. Ma, and Y-X. Huo. 2020. "Genetic Engineering of Non-Native Hosts for 1-Butanol Production and Its Challenges: A Review." *Microbial Cell Factories* 19(1): 79.

Rapoport, D. C. 2004. "The Four Waves of Modern Terrorism." In *Attacking Terrorism, Elements of a Grand Strategy*, edited by Audrey Kurth Cronin and James M. Ludes, 46–73. https://press.georgetown.edu/Book/Attacking-Terrorism.

Reymond, J. -L. 2015. "The Chemical Space Project." *Accounts of Chemical Research* 48(3): 722–730. https://doi.org/10.1021/ar500432k.

Santos, C, et al. 2019. "Characterizing Chemical Terrorism Incidents Collected by the Global Terrorism Database, 1970–2015." https://www.cambridge.org/core/journals/prehospital-and-disaster-medicine/article/abs/characterizing-chemical-terrorism-incidents-collected-by-the-global-terrorism-database-19702015/DDC014FA265072B42234F86C4D337262.

Schuurman, B. 2020. "Research on Terrorism, 2007–2016: A Review of Data, Methods, and Authorship.," *Terrorism and Political Violence* 32(5): 1011–1026. https://doi.org/10.1080/09546553.2018.1439023.

Smithson, A. E., and L.-A. Levy. 2000. "Chapter 3—Rethinking the Lessons of Tokyo" (pdf). Ataxia: The Chemical and Biological Terrorism Threat and the U.S. Response (Report). Henry L. Stimson Centre. 91–95, 100. Report No. 35. Archived from the original on December 25, 2014.

START (National Consortium for the Study of Terrorism and Responses to Terrorism). 2022. Global Terrorism Database 1970–2020 [Data File]. https://www.start.umd.edu/gtd.

Teal, C. 2023. "Florida City Water Cyber Incident Allegedly Caused by Employee Error." *Route Fifty*, March 21, 2023. https://gcn.com/cybersecurity/2023/03/florida-city-water-cyber-incident-allegedly-caused-employee-error/384267.

Tennenbaum, M., and M. E. Kosal. 2021. "The Interplay Between Frugal Science and Chemical and Biological Weapons: Investigating the Proliferation Risks of Technology Intended for Humanitarian, Disaster Response, and International Development Efforts.," In *Weapons Technology Proliferation: Diplomatic, Information, Military, Economic Approaches to Proliferation*, edited by M. E. Kosal, 153–204. Springer Academic Publishers, August 2021. https://www.springer.com/us/book/9783030736545.

Tucker, J. B. ed. 2000. *Toxic Terror: Assessing Terrorist Use of Chemical and Biological Weapons*. MIT Press.

Tucker, J. B. 2006. *War of Nerves: Chemical Warfare from World War I to Al-Qaeda*. Anchor.

Uddin, M. I., N. Zada, F. Aziz, Y. Saeed, A. Zeb, S. A. A. Shah, M. A. Al-Khasawneh, and M. Mahmoud. 2015. "Prediction of Future Terrorist Activities Using Deep Neural Networks." *Complexity* 2020(1): 16, Article ID 1373087. https://doi.org/10.1155/2020/1373087.

Urbina, F., F. Lentzos, C. Invernizzi, and S. Ekins. 2022. "Dual Use of Artificial-Intelligence-Powered Drug Discovery." *Nature Machine Intelligence* 4(3): 189–191. https://doi.org/10.1038/s42256-022-00465-9.

Vasquez, C. 2023. "Did Someone Really Hack into the Oldsmar, Florida, Water Treatment Plant? New Details Suggest Maybe Not." *CyberScoop*, April 10, 2023. cyberscoop.com/water-oldsmar-incident-cyberattack.

Vilkhovoy, M., A. Adhikari, S. Vadhin, and J. D. Varner. 2020. "The Evolution of Cell Free Biomanufacturing," *Processes* 8(6): 675–694.

Yan, Q., and B. F. Pfleger. 2020. "Revisiting Metabolic Engineering Strategies for Microbial Synthesis of Oleochemicals.," *Metabolic Engineering* 58: 35–46.

Zong, J., and J. Yue. 2022. "Continuous Solid Particle Flow in Microreactors for Efficient Chemical Conversion." *Industrial & Engineering Chemistry Research* 61(19): 6269–6291. https://doi.org/10.1021/acs.iecr.2c00473.

3

Evaluation of Strategies

3.1 OVERVIEW OF STRATEGIES ASSESSED

In accordance with the Statement of Task (SOT), the committee assessed a variety of strategic documents, summarized in Table 3-1. Important to note is that while these documents all provide strategic guidance, only some are strategies, while others are policy, public guidance documents, mission overviews, frameworks, or joint doctrine. For the purpose of this report, the committee will refer to these documents as the "strategy documents." Each of these types of documents is meant to be used in concert, but they may serve different purposes:

- **Joint Doctrine** presents "fundamental principles that guide the employment of U.S. military forces in coordinated and integrated action toward a common objective." It promotes a common perspective from which to plan, train, and conduct military operations. It represents what is taught, believed, and advocated as to what is right (i.e., what works best). "It provides distilled insights and wisdom gained from employing the military instrument of national power in operations to achieve national objectives" (Joint Chiefs of Staff, n.d.). As such, *doctrinal documents inform how strategy can be implemented in accordance with any relevant policy.*

    Joint Doctrine is intended only to be revised when geopolitical circumstances or U.S. policy change significantly enough that a change to how U.S. military forces are employed has been deemed necessary.
- **Strategy** is a comprehensive plan that outlines the specific actions and decisions an organization should make in situations that may occur in the future in order to achieve a desired outcome. According to Joint Doctrine Note 2-19, at the national level, the strategy's ultimate goal is to *"achieve policy objectives by*

maintaining or modifying elements of the strategic environment to serve the interest" outlined in U.S. policy (O'Donohue, 2019).

• **Policy** is a set of guiding principles that outline organizational rules for activities that are repetitive in nature. In short, it outlines what should and should not be done by the organization. According to Joint Doctrine Note 2-19, at the national level, *policy documents summarize "the positions of governments and others cooperating, competing, or waging war in a complex environment"* (O'Donohue, 2019).

While many strategy documents were reviewed and considered by the committee (see Appendix A, Table A-1 for a complete list), the committee chose to focus its efforts on the eight documents listed in Table 3-1. As a note, the *2023 DoD Strategy for Countering Weapons of Mass Destruction* was not subject to the Committee's evaluation methodology.

3.2 METHODOLOGY OF ASSESSMENT

Well-reasoned and intentioned efforts to ensure accountability, critically important in the context of the use of national security, public trust, and good use of tax-payers money, often drive requirements for metrics (Muller, 2018). One of the biggest challenges can be expressed using Goodhart's law: "When a measure becomes a target, it ceases to be a good measure" (Strathern, 1997). Ensuring that metrics are meaningful (rather than trivial, political, idiosyncratic, or less than meaningful) is hard. The classic case in national security comes from metrics on success put in place by the DoD for the military services during the Vietnam War (Daddis, 2012). Previous high-level efforts to define "broad-based objective criteria when evaluating progress in the nation's efforts to combat terrorism" have noted challenges:

> A common pitfall of governments seeking to demonstrate success in anti-terrorist measures is overreliance on quantitative indicators, particularly those which may correlate with progress but not accurately measure it, such as the amount of money spent on anti-terror efforts (Perl, 2007, Pg. 3).

This committee has aimed to create and employ an objective, repeatable methodology for assessing the effectiveness of strategies that incorporates metrics and also seeks to ensure that they are meaningful and comprehensive, considering the whole rather than focusing on single metrics or limited cases.

To address the adequacy of strategies to identify, prevent/counter, and respond to chemical terrorism, the committee adopted a methodology that employs a systematic approach to evaluating relevant documents. Appendix D provides the complete rubric. Several documents ranging from national level to individual agencies strategies were evaluated in detail. These were made publicly available within the last 10 years since the formation of this consensus study committee. The adequacy of these strategies was examined through three different lenses—identity, prevention/countering, and response. Before reviewing the documents, a list of characteristics that define an adequate strategy

TABLE 3-1 Key Strategic Documents

Document
Office of the Press Secretary, 2018. "National Strategy for Countering Weapons of Mass Destruction Terrorism," available at: https://www.hsdl.org/?view&did=819382.
Office of the Under Secretary of Defense for Policy, 2017. "DoD Directive 2060.02: DoD Countering Weapons of Mass Destruction (WMD) Policy," available at: https://www.esd.whs.mil/Portals/54/Documents/DD/issuances/dodd/206002_dodd_2017.pdf.
Joint Chiefs of Staff, 2019. "Joint Publication 3-40: Joint Countering Weapons of Mass Destruction," 2021, available at: https://www.jcs.mil/Portals/36/Documents/Doctrine/pubs/jp3_40.pdf.
Joint Chiefs of Staff, 2016. "Joint Publication 3-41: Chemical, Biological, Radiological, and Nuclear Response," available at: https://www.jcs.mil/Portals/36/Documents/Doctrine/pubs/jp3_41.pdf.
Office of the Deputy Assistant Secretary of Defense for Chemical and Biological Defense, 2020. "Chemical and Biological Defense Program (CBDP) Enterprise Strategy."
U.S. Department of Defense, 2022. "2022 National Defense Strategy of the United States of America," available at: https://media.defense.gov/2022/Oct/27/2003103845/-1/-1/1/2022-NATIONAL-DEFENSE-STRATEGY-NPR-MDR.PDF
The White House. 2022. National Security Strategy. https://www.whitehouse.gov/wp-content/uploads/2022/10/Biden-Harris-Administrations-National-Security-Strategy-10.2022.pdf
U.S. Department of Homeland Security, 2019. "Department of Homeland Security Chemical Defense Strategy."
Department of Homeland Security. 2008. "National Response Framework." https://www.fema.gov/pdf/emergency/nrf/about_nrf.pdf

was created for each category. Then, each evaluation was measured against the same rubric, which allows for a qualitative yet consistent approach to assessing the strategies across the three major groups. In cases where the strategy does not address the topic, the document was not evaluated.

The rubric addresses the following questions:

1. Does a genuine strategy exist, and if so, to what extent is it coherent?
2. Does the strategy sufficiently meet the chemical threat over the required timeframe of interest? If so:
   a. To what extent do the goal(s) collectively encompass the level and type of threat that is likely to emerge in that timeframe?
   b. What policies, plans, and resource allocations are enabling the goals to be achieved?
3. How feasible is the strategy concerning statutes, fiscal, and politics?

The cohesiveness of the strategy was answered by initially scanning for two key variables: 1) stated goals related to either identify, prevent/counter, or response and 2) clear definition(s) of success for when a goal is achieved. The validity of each stated goal was further assessed by whether the USG has any policies, plans, or resource allocations available to address them. Each identified goal and its corresponding policies, plans, and resource allocations were also verified for explicit documentation in the strategy and consistency between them. The effectiveness of the goals was also assessed on how well they can respond to a chemical threat that is likely to occur within a relevant timeframe and possibly beyond the nature of the threat. Policies, resource allocations, plans, and their ability to enable the goals were also measured against the timeframe. The legal, fiscal, and political feasibility of implementing various aspects of the strategies were also explored. Finally, the overall adequacy of the strategy was assessed using a qualitative ranking system that ranged from exceeding to inadequate, with partially inadequate, partially adequate, and adequate as other choices within the spectrum. While the committee recognized that this methodology is incomplete and the sample size of documents examined is small, the rubric employed in this assessment was useful for providing a centralized and consistent platform to evaluate the three major categories and communicate the respective findings, conclusions, and recommendations. The committee also found this method valuable for extrapolating findings relevant to de facto strategies.

Chapters, 4, 5, 6, and 7 will discuss in detail the general findings from the strategy assessments and will also include any technical, policy, or resource gaps found that, if included could strengthen parts of the strategies. In addition to the results from this method, the chapters will provide other types of evidence from various sources (e.g., briefing presentations from federal agencies, literature, and congressional hearings).

REFERENCES

Daddis, G. A. 2012. "The Problem of Metrics: Assessing Progress and Effectiveness in the Vietnam War." *War in History* 19(1): 73–98. https://doi.org/10.1177/0968344511422312.

DHS (Department of Homeland Security). 2019. National Response Framework. 4th ed. https://www.fema.gov/sites/default/files/2020-04/NRF_FINALApproved_2011028.pdf.

DoD (U.S. Department of Defense). 2022. Fact Sheet: 2022 National Defense Strategy. https://media.defense.gov/2022/Mar/28/2002964702/-1/-1/1/nds-fact-sheet.pdf.

Joint Chiefs of Staff. n.d. Joint Doctrine Publications. https://www.jcs.mil/Doctrine/Joint-Doctrine-Pubs.

Muller, J. Z. 2018. *The Tyranny of Metrics*. Princeton, NJ: Princeton University Press.

O'Donohue, D. 2019. Joint Doctrine Note 2-19. Strategy. https://www.jcs.mil/Portals/36/Documents/Doctrine/jdn_jg/jdn2_19.pdf.

Perl, R. 2007. Combating Terrorism: The Challenge of Measuring Effectiveness. CRS Report, March 12, 2007. https://apps.dtic.mil/sti/citations/ADA466584.

Strathern, M. 1997. "Improving Ratings: Audit in the British University System." *European Review* 5(3): 305–321.

The White House. 2022. National Security Strategy. https://www.whitehouse.gov/wp-content/uploads/2022/10/Biden-Harris-Administrations-National-Security-Strategy-10.2022.pdf.

4

Adequacy of Strategies to Identify Chemical Threats

Summary of Key Findings, Conclusions, and Recommendations

FINDING 4-1: Most federal agencies surveyed by the committee acknowledge that overall terrorists seeking to perpetrate chemical attacks tend to opportunistically misuse traditional classes of chemicals, primarily toxic industrial chemicals and toxic industrial materials.

FINDING 4-2: The federal agencies that briefed the committee indicated that the total number of potential chemical threats—whether existing, emerging, or yet to be designed—that can or could be used for weapons of mass destruction is vast and expanding.

FINDING 4-3: The agencies surveyed are broadly aware of each other's efforts to identify, prevent, counter, and respond to chemical threats, but express concern that information-sharing and coordination across relevant agencies is incomplete.

CONCLUSION 4-4a: It is impossible to identify, prevent, or counter every threat. Overall, the majority of publicly reported domestic plots did not come to fruition between the 1970s through the mid-2010s for a number of reasons.

CONCLUSION 4-4b: The Federal Bureau of Investigation (FBI) and partner law enforcement and intelligence communities (IC) have been effective in identifying and interdicting the majority of domestic terrorist attacks involving chemical materials, which have typically employed conventional toxic industrial chemicals rather than traditional chemical warfare agents, such as sarin. While the FBI has been effective, approaches to identifying chemical threats could be strengthened

continued

Summary continued

using a multilens approach from several different agencies that emphasizes augmented communication and coordination between local and state enforcement and IC. In addition, this area would greatly benefit from increased coordination between the IC and technical experts (particularly those with specific expertise in the areas of terrorist motivation and psychology). For example, FBI antichemical terrorism resources focused on identification could be evaluated in the context of current identification strategies employed by other agencies.

RECOMMENDATION 4-4: Existing intelligence community programs should actively seek and incorporate new approaches to identify existing chemical threats (traditional and improvised) and potential emerging threats by terrorist groups. In developing new approaches, program managers should develop strategies that look beyond the traditional terrorism suspects and that augment and leverage skill sets of the U.S. Government agencies. For example, scholars of political psychology could work with chemical terrorism experts to create a holistic approach of identifying chemical terrorist groups or similar violent actors outside the traditional suspects. The threat assessments should be improved by reflecting the current times and demographics.

FINDING 4-5: The shift to great power competition may change the nature of the threat for new chemical attacks, in that chemical agents, other materials, technology, and expertise may migrate from state actors that engage in either defensive or offensive activities to violent extremist organizations (VEOs). These events could enable VEOs to conduct more sophisticated attacks, with agents and/or with means of delivery not otherwise accessible to them.

RECOMMENDATION 4-5: The National Counter Terrorism Center (NCTC), Department of Defense (DoD), Department of Homeland Security (DHS), and State Department should review current identification approaches to determine whether shifts in emphasis are required as a result of expanded and augmented VEOs and terrorist capability resulting from the potential migration of chemical agents, other materials, technology, and expertise from state actors to VEOs.

CONCLUSION 4-6: It is unclear if the tactical readiness to implement the reviewed strategies is occurring at the necessary pace to respond to an act of chemical terrorism. Additionally, the shift in strategic focus to great power competition (GPC) may lead to reduced resources for countering acts of terrorism employing weapons of mass destruction that are perpetrated by VEOs, and may impede tactical readiness against chemical terrorist threats, leading to increased risk.

RECOMMENDATION 4-6: The United States Government (USG) should ensure that the identification of chemical terrorism threats is explicitly included in ongoing and future strategies. Chemical terrorism threats should be considered distinct from nuclear nonproliferation, identification of state-based offensive chemical programs, and traditional (non-nuclear-biological-chemical) terrorism.

Effective strategic communications can be compared to an orchestra producing harmony (see Figure 4-1). Best practices include the empowerment of a single individual—the conductor, who coordinates and integrates the various instruments—all of which retain their unique sound and specialty while communicating more effectively in concert. Further, the conductor must continuously adapt their interpretation of the score based on stakeholder feedback:

> *The panoply of U.S. force actions must be synchronized across the operational battlespace to the extent possible so as not to conflict with statements made in communications at every level from President to the soldier, sailor, marine, or airman on the street* (Rand Corporation 2007).

The committee's discussions with agency representatives (see Appendix A) highlighted important gaps in the area of identifying, communicating, and responding to chemical threats that should be addressed to enable timely response to real-world weapons of mass destruction terrorism (WMDT) incidents involving chemical threats.

One of the most significant challenges associated with a successful implementation of strategies to counter WMDT chemical threats is a smooth coordination of the various activities to ensure actionable threat awareness. Using the orchestration analogy described earlier to actionably identify WMDT chemical threats, the National Security Council (NSC) composes the score—specifically, the Global Chemical Deterrence Framework and NCTC plays the role of conductor, convening representatives from across the IC as well as other stakeholder agencies responsible for preventing, countering, and responding to carry out lines of effort outlined in the score (the Global Chemical Deterrence Framework).[1] As the composer, the NSC plays a significant role in leading planning, and documenting specific actions to be taken by each stakeholder agency, promulgating the guidance, and integrating mature capabilities into strategic guidance to enable chemical threat recognition and response at timescales of relevance. As the conductor, NCTC then interprets the NSC's guidance (the Global Deterrence Framework and other strategic guidance) and uses it to convene other IC agencies to assess threats and communicate findings to other stakeholder agencies responsible for preventing, countering, and responding in a timely enough manner to enable their mission success. The various IC agencies and other stakeholders, as fellow members of the orchestra, invest in needed research and integrate mature research into operational use to enhance their abilities to play their parts (e.g., conduct their respective missions). If the strategic guidance is implemented effectively, the IC's synchronized activities will enable the USG to be nimbler in the face of evolving threats, in turn facilitating more accurate and actionable identification. Therefore, **it is important that the NSC institutionalizes the information-sharing efforts being conducted by NCTC and that lines of effort be adjusted as threats evolve.**

Prioritization of whether a chemical should be considered a threat largely occurs through coordination within interagency working groups. These activities have grown into robust avenues for information exchange that may yield benefits for successfully identifying and countering WMDT chemical threats as well as collaboratively adjusting lines of effort as threats develop. The committee's discussions with the agencies involved with

---

[1] DoD JP-40, Appendix A.

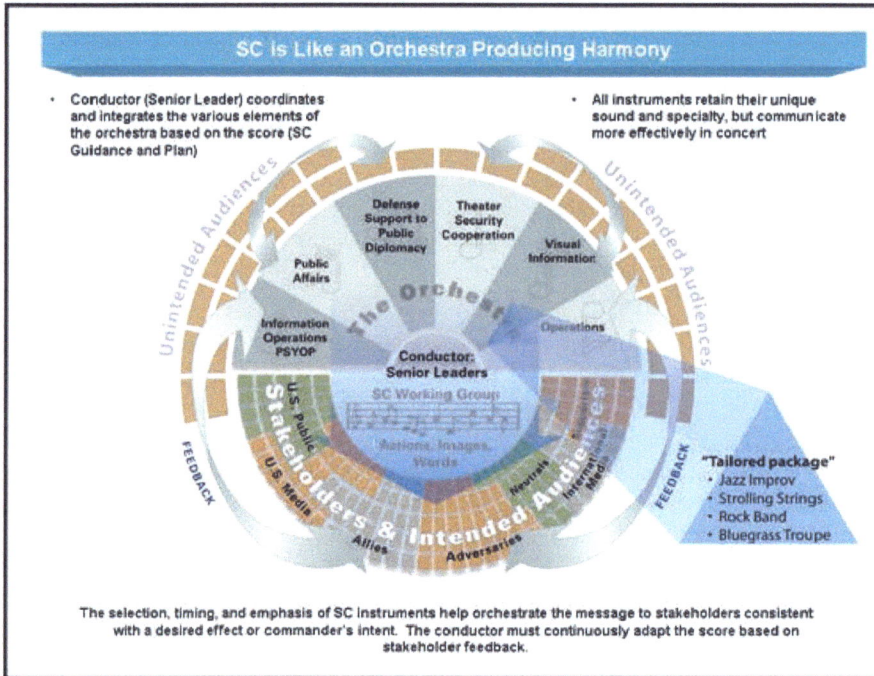

FIGURE 4-1 Strategic Communication (SC) is like an orchestra producing harmony.
SOURCE: Commander's Handbook for SC and Communication Strategy (Ver2.0 U.S. Joint
Forces Command, 2009).

research, development, testing, and evaluation (RDT&E) focused on activities to recognize and respond to chemical threats. Box 4-1 outlines three major themes that were heard.

Based on briefings and follow-on discussions conducted by the committee, opportunities for improvement in communications at the **tactical, operational,** and **strategic** levels are needed for establishing, or updating, and carrying out strategies assessed in support of this study (see Figure 4-2). In terms of strategic communication, the tactical level focuses on direct information engagement between agencies and other entities; the operational level deals with commanders' and agency directors' communications strategies both for staff in their organizations and with external stakeholders; and the strategic level deals with national-level strategic communications. While the committee was made aware of some best practices, these appeared to be conducted on an as needed basis and were largely driven by individual personalities or relationships within or between agencies.

Tactical: Briefings provided by a few of the IC agencies highlighted the need to engage more stakeholders in relevant communication mechanisms, such as working groups and issues-focused events. In addition, the committee believes that the benefit of including more technical experts—from Department of Energy (DOE) laboratories, private laboratories, and universities—in interagency discussions beyond those person-

---

BOX 4-1
Three Themes Committee Heard from
Information-Gathering Meetings

1. Important work is being undertaken to identify chemical terrorism threats and to develop capabilities to respond.
2. The successes communicated to the committee are at the basic research level and require further optimization before these approaches can be integrated into strategy and tactical operations.
3. It is not clear that the briefers were fully cognizant of efforts and key capabilities resident in other agencies, or of the need to ensure that the capabilities being developed in one agency are broadly available to other agencies.

---

nel from DoD laboratories far outweighs the potential security concerns. The negative strategic impact of an undetected Skripal poisoning-like scenario (see Appendix E) on the homeland is an example of why it is important to include all relevant experts surrounding the issue at hand. Certain ad hoc interagency working groups have done this successfully and could serve as a standard for routine engagement.

Operational: The committee heard several briefings that show excellent coordination between DoD's research arms (Defense Threat Reduction Agency [DTRA], The Joint Program Executive Office [JPEO], U.S. Army Combat Capabilities Development Command [CCBC], Chemical Biological Center [CBC]) when it comes to preparing and equipping the warfighter to identify, prevent, counter, or respond to WMDT chemical threats. However, it is not clear that this level of coordination exists between these DoD entities and other research institutions (e.g., DOE laboratories, private institutions, or universities, where their expertise is relevant) or for preparing and equipping warfighters' civilian counterparts to identify, prevent, counter, or respond to WMDT attacks on the homeland involving chemicals.

The Federal Bureau of Investigation (FBI) has strong functioning relationships with the DHS and maintains personnel within the DHS WMD organizations, which serves as an example of an organization that has strong operational communications (Savage, 2022). The FBI also maintains relationships with the National Laboratory community to anticipate future chemical threats. Communications between the FBI and state and local authorities is achieved through the Threat Credibility Office which is staffed 24/7; this office functions to convene experts throughout the duration of an event. The WMD coordinators are also on call, and they have relationships with first responders and the National Guard. Interaction of these entities with the National Guard is exercised frequently through the Threat Credibility Evaluation (TCE) process, which includes the Livewire program. Annual exercises involving full-field response to a simulated WMD incident are conducted. The FBI maintains robust online surveillance, the details of which are at a higher level of classification (Rotolo, 2022).

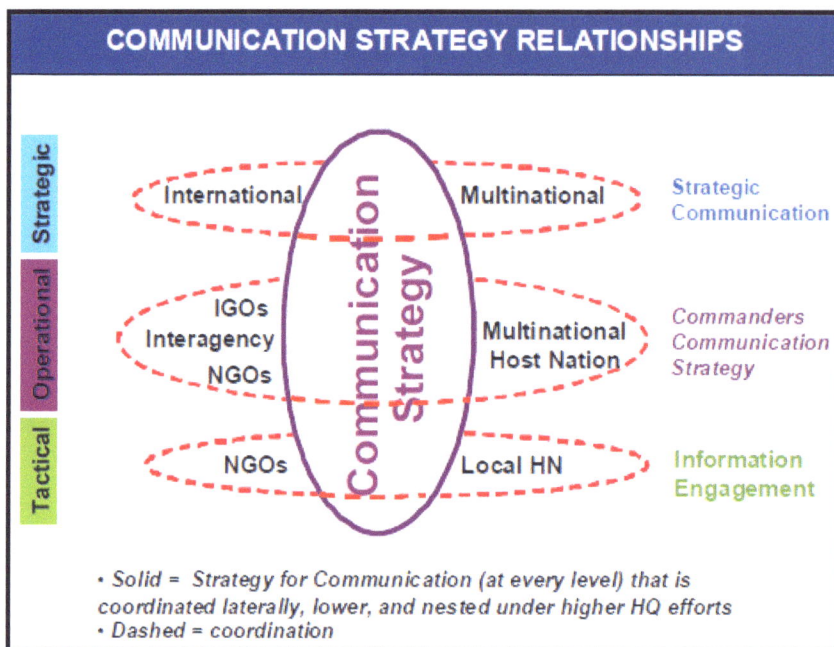

FIGURE 4-2 Communication Strategy Relationships. Ideally, communication strategy at every level is coordinated laterally, lower, and nested under higher headquarters (HQ) efforts. SOURCE: Ver2.0 (U.S. Joint Forces Command, 2009).

Strategic: A renewed focus on strategic communication is needed—either to counter the narratives put forward by adversaries or to deter attacks by communicating capabilities that lead adversaries to conclude that WMDT involving chemicals is unlikely to yield the desired success. Efforts in 2022 undertaken by the DTRA-Cooperative Threat Reduction (DTRA-CTR) program illustrated a successful example of strategic communication. DTRA-CTR identified a potential biological threat imposed through disinformation;[2] specifically, the organization proactively communicated the goals and successes of its partnerships with public health laboratories in Ukraine to counter the narrative put forth by popular media personalities and others, where the disinformation suggested nefarious USG involvement to produce biological weapons-related laboratories (NPR, 2022). The level of proactive communication recently demonstrated by the DTRA-CTR program and its leadership in countering disinformation could be emulated and adopted into more routine practice by other relevant agencies and programs.

Like conducting an orchestra, successful communication across relevant agencies is essential for identifying chemical threats. All stakeholders involved should receive the intended message clearly and be able to cohesively relay the information to others.

---

[2] Disinformation is deliberately crafted to mislead, harm, or manipulate a person, social group, organization, or country (CIS, n.d.).

Nested within this paradigm are tactical, operational, and strategic communications; all functions necessary to manage, implement, and communicate the directions for identifying, preventing, and responding to immediate and potential chemical threats. The following discussion considers these aspects with respect to "identify" within the Committee's evaluation of the strategy documents.

4.1 ANALYSIS OF STRATEGIES TO "IDENTIFY" WMDT CHEMICAL THREATS

As outlined in the description and analysis of the Baseline Threat in Chapter 2, while the number of terrorist incidents involving the use of WMD over the last two decades is low in comparison to terror attacks using conventional weapons, WMD-involved terror attacks that employed chemicals make up the largest percentage of such attacks. Therefore, in assessing existing strategies' sufficiency to actionably identify chemical terrorism threats, it is important to consider the following two points:

- The occurrence or nonoccurrence of terror attacks involving chemicals is not a direct indication that the United States was or was not successful in identifying a particular threat.
- The existence or absence of these chemical terror attacks does not imply a strategy's sufficiency, or that the strategy has been successfully or properly implemented.

Significantly more detailed analysis beyond the scope of this study would be required in order to make such an assessment that adequately evaluates these two parameters. The committee recognizes their assessment is necessarily based on the information available to the committee and within the constraint of the time and resources allotted for the study; thus, it is also important to recognize it is impossible to know beforehand every possible chemical threat. This level of information is not a criterion for success in this analysis.

Box 4-2 lists the documents that were analyzed for "identify" against the methodology described in Chapter 3. The committee recognizes that this subset of documents, while selected to be inclusive and representative of major programs across the USG, may have limitations; however, they serve as an appropriate representation of strategies put forth by key agencies that are able to implement actions for successfully identifying chemical terrorism threats.

4.1.1 Committee Definition of Adequacy: Identify

The committee views that a successful strategy to identify chemical terrorism threats focuses on robust information-sharing regarding the following:

1. Chemicals that may be used in an attack—both known chemical weapon agents, toxic industrial chemicals/toxic industrial materials (TICs/TIMs), and lesser known emerging agents;

BOX 4-2
Strategy Documents Reviewed by
Committee for "Identify" Analysis

1. White House. (2018). National Strategy for Countering Weapons of Mass Destruction Terrorism.
2. Department of Homeland Security. (2019). Department of Homeland Chemical Defense Strategy.
3. Office of the Deputy Assistant Secretary of Defense for Chemical and Biological Defense. (2020). Enterprise Strategy.
4. Department of Defense. (2017). National Security Strategy of the United States of America, National Strategy for Countering Weapons of Mass Destruction Terrorism.
5. Joint Chiefs of Staff, Joint Publication 3-40. (2019, validated 2021). Joint Countering Weapons of Mass Destruction.

---

2.   Threat actors who may use or pursue chemicals for use in WMDT attacks; and
3.   Entities that may support or sponsor chemical attacks or terrorism.

The committee's analysis considered all three of these aspects. Successful "identify" strategies focus on developing actionable information to facilitate preventing, countering, and responding to the identified threats. A strategy had to be determined to adequately address each (prevent, counter, respond) to be judged "adequate" in the area of identify.

4.1.2 Clearly Defined Ends Are Adequate, but
Ways and Means Are Not Apparent

The strategy documents reviewed by the committee all have clearly stated goals that include an explicit definition of success. In particular, the Chemical and Biological Defense Program (CBDP) Enterprise Strategy provides an explanatory paragraph describing what the organization views as success. Using a standard description of ways to assess strategy (Deibel, 2007), we found that in the CBDP strategy, a "**desired end**" was applied to each stated goal, along with detailed descriptions of "**ways**" (e.g., how) the goals were to be achieved. However, explicit descriptions of available resources to implement the strategies, "**means**," were absent from the documents' discussion.

The committee was able to glean some information regarding "means" during briefings with representatives from DTRA, DHS, Biomedical Advanced Research and Development Authority (BARDA), U.S. Army Combat Capabilities Development Command Chemical Biological Center (DevCom CBC), and the State Department (see Table A2 in Appendix A). However, DoD CBDP's ability to orchestrate other agencies' use of the means available to them remains unclear. Specifically, under the CBDP enterprise strategy cross-cutting goal, "Drive Innovation," stated objectives include broadening and strengthening relationships within DoD and the Enterprise and building internal and external partnerships with the IC to ensure intelligence support to Enterprise research. To accomplish these objectives, CBDP will increase the frequency and quality of engagements with partners seeking complementary solutions, such as Public Health agencies while leveraging the best practices in the biopharmaceutical industry to speed RDA and regulatory approval of vital MCMs and focus partnerships with American industry to help align private sector research, development, and acquisition (RDA) to national security priorities. However, CBDP cannot achieve the intended increased engagement on its own; it must be met by corresponding prioritization and enablement of engagement on the part of other agencies. In the committee's discussion with DHS Countering Weapons of Mass Destruction Office (DHS CWMD). The speakers observed that stakeholders responsible for addressing known or emerging chemical threats lack both the authority and sufficient "means" in the form of personnel resources or funding to act with an adequate level of completeness or timeliness. This insufficiency creates a challenge for DHS CWMD to effectively deploy their charge of coordinating and ensuring information-sharing across the whole of government, private, industry, and state, local, tribal, and territorial (SLTT). The DHS CWMD strategy document demonstrated similar observations. The fourth goal in the strategy related to threat identification stated, "collaborate with SLTT governments, private sector partners and other key stakeholders to prioritize and share timely, accurate and actionable information," (DHS, 2019, Pg. ii) however, the goal did not specify available resources ("means") or levers available to encourage the occurrence ("ways") of this type of collaboration. The fact that both the CBDP Enterprise Strategy and the DHS CWMD Strategy include goals associated with improving collaboration and coordination may indicate that CBDP and DHS CWMD are working at cross-purposes.

Based on these observations, the committee judges the strategies to be **partially adequate.** In the committee's review of the strategy documents and through briefings and discussions with representatives from DHS, DTRA, DevCom CBC, State Department, NCTC, and BARDA, clearly defined lines of effort were evident, and the representatives were able to describe both ways in which they are implementing the strategies and means being used to do so. The committee judged that the specific goals and supporting lines of effort outlined in the strategy documents are both appropriate for accomplishing the desired ends and are coherent overall. Although, briefers repeatedly acknowledged the reality that it is impossible to identify every single threat and further, that the level of resourcing is insufficient to be fully successful. Additionally, none of the strategies addressed how it is intended for them individually or collectively to sustain enduring success over time, given the reality of shifting political priorities and the ebbs and flows of funding that inevitably result from such shifts.

4.1.3 Roles of TICs TIMs in Increasing Risk of Chemical Terror Attacks

The threat of terrorism involving TICs/TIMs has been detailed in chapter 2. The committee reviewed how the strategy documents addressed the identification of threats associated with TICs/TIMs, in order to test the soundness of the "identify" analysis.

DHS Defense Strategy emphasizes the identification and mitigation of threats originating from TICs and TIMs opportunistically used in a chemical attack by VEOs. The DHS Strategic Plan (DHS, 2022) states that "chemical materials and technologies with dual-use capabilities are more accessible throughout the global market," (Pg. 15) and that the proliferation of information and technologies provides augmented opportunities for rogue nations and nonstate actors to develop, acquire, and use WMD.

DHS Chemical Defense Strategy (DHS, 2019) notes that state and nonstate actors have deployed TICs and TIMs in a variety of offensive uses. DHS acknowledges that manufacturing, storage, and transportation infrastructure pose a danger as sources of a release. The DHS Chemical Defense Strategy acknowledges that identification of these threats is complicated by the reality that the TICs have legitimate industrial, agricultural, or pharmaceutical applications and that production may be concealed "within industrial or agricultural production facilities, and academic or pharmaceutical labs" (DHS, 2019, Pg. 5). A specific DHS objective focuses on both state and nonstate actors plotting or perpetrating incidents involving the chemical industry so that chemical incident risks and adversary capabilities that might employ TICs and TIMs are understood. Further, DHS is engaged with characterizing and forecasting chemical risks specific to geographic and economic sectors as potential terrorism targets, which clearly indicates TICs and TIMs as a major source of concern.

DHS's concern regarding TICs and TIMs is further substantiated by the National Infrastructure Protection Plan (NIPP) (DHS, 2013), which emphasizes the assessment and analysis of risks derived from storage, manufacture, and transportation of chemicals. Concern over the use of TICs and TIMs is also reflected in Chemical Facility Anti-Terrorism Standards (CFATS), which developed standards for chemical safety and security at facilities where chemicals are stored and manufactured. The CFATS program included a standard focused on sabotage, which would include attacks by VEOs. These observations indicate that DHS considered incidents where TICs and TIMs would be employed to be significant threats.

In contrast, the DoD strategy does not emphasize threats involving VEOs utilizing TICs and TIMs. DoD strategy notes that response to domestic events is to be led by the FBI for all terrorist-related incidents and threats (JCS, 2019); however, with regard to chemical threats, the DoD strategy acknowledges that the Department may play a supporting role in response (DoD, 2014). Furthermore, the DoD Task Force on Deterring, Preventing, and Responding to the Threat or Use of Weapons of Mass Destruction noted that the threat from adversaries, both military and civilian, was growing and that it was difficult to detect before the event (DSB, 2018). The DoD report specifically noted that new experimentation in the uses of TICs was on the rise and that there was reasonable willingness of organizations to use these.

> FINDING 4-1: Most federal agencies surveyed by the committee acknowledge that overall, terrorists seeking to perpetrate chemical attacks tend to opportunistically misuse traditional classes of chemicals, primarily toxic industrial chemicals and toxic industrial materials.

4.1.4 Increasing Diversity of Chemical Threats

The identification of chemical threats has become more challenging, and the difficulty in chemical threat identification is expected to increase. The array of chemical substances and materials that have been employed as chemical agents has increased, a reality that is acknowledged by the DoD strategy documents (JCS, 2019; The White House, 2018). Furthermore, the number of chemicals with potential for use as agents will certainly continue to grow. In addition, the identification of threats is exacerbated by an increased diversity of actors and the dual-use nature of related technology and expertise (Trump, 2017). Further complications are anticipated because of breakthroughs in chemistry resulting in the generation of deadlier chemical agents, such as fourth-generation agents (JCS, 2019). Nontraditional agents, such as pharmaceutical-based agents (PBAs) (DOS, 2022), and nerve agents expand the range of substances that could be considered as potential threats (Hersman et al., 2019). The likelihood of these threats materializing or being identified could be augmented by increasing availabilities of emerging science and technologies (e.g., advanced material science, AI/ML, small-scale reactors) as discussed in Chapter 2.

> FINDING 4-2: The federal agencies that briefed the committee indicated that the total number of potential chemical threats—whether existing, emerging, or yet to be designed—that can or could be used for weapons of mass destruction is vast and expanding.

4.1.5 Cross-Agency Communication

The "identify" function is potentially problematic to the countering of weapons of mass destruction and terrorism (CWMDT) endeavor because there are multiple agencies involved, in terms of chemicals that could be used as agents (whether current or emerging chemical weapons agents (CWAs), or TICs and TIMs), targets, and prospective perpetrators. Agencies conducting identification activities may not be those who would be involved with deterrence or interdiction. Therefore, interagency communication is of paramount importance. JP 3-40 states:

> *CWMD requires unity of effort, which results in a coordinated response of combined capabilities of the USG. Coordination between DoD and other USG departments and agencies is critical to the success of CWMD operations against the global WMD threat* (JCS, 2019, A-6).

As mentioned earlier, the NSC Staff has the responsibility of overseeing lines of communications between USG departments and agencies involved in CWMDT activities with the objective being to leverage all instruments of national power—the orchestrator (JCS, 2019).

The subsequent sections (4.2.5a and 4.2.5b) discuss two major challenges observed by the committee in their analysis of the strategies. The first challenge involves disclosing pertinent information and balancing the risks involved in identifying and communicating the chemical threat. Second, understanding the different coordination roles and responsibilities of each agency remains a challenge between relevant entities. The analysis of the strategy documents allowed the committee to unpack specific details underpinning these issues.

4.1.5a Protecting Sensitive Information and Ensuring Adequate Identification of Chemical Terror Threats

DHS strategy clearly emphasizes communication, stating that the agency will collaborate with SLTT, the private sector, and others for prioritizing and sharing timely accurate and actionable information (DHS, 2019). Information-sharing with federal agencies and first responders is important, however, a challenge is to "closely scrutinize classification levels to achieve the broadest distribution of information, while protecting sensitive information." (DHS, 2019, Pg. 7). Information-sharing is also emphasized in the National Infrastructure Protection Plan (NIPP, a DHS document) (DHS, 2013a). Specifically, the NIPP includes a detailed section specifying the establishment of Sector Coordinating Councils, which are to enable strategic communication and coordination between the private sector and government in response to emerging threats, or response and recovery operations. In parallel, the NIPP includes government coordinating councils to ensure cross-jurisdictional coordination. This is to ensure information-sharing across sectors and to promote public-private information-sharing. The NIPP establishes four cross-sector councils for the purpose of planning: these address (1) critical infrastructure, (2) Federal Senior Leadership, (3) SLTT government, and (4) the regional consortium Coordinating Council. A key concept in the NIPP is that the document "integrates efforts by all levels of government, private, and non-profit sectors by providing an inclusive partnership framework and recognizing unique expertise and capabilities each participant brings to the national effort." (DHS, 2013b) The NIPP also establishes the National Infrastructure Coordinating Center (NICC) and the National Operations Center, which are focused on cross-agency and public/private sector communication and coordination.

4.1.5b Recognizing Roles and Responsibilities of Agencies and Programs

DoD strategy acknowledges that the State Department is the "USG lead agency for CWMD operations abroad" and that DoD has a supporting role (JCS, 2019, A-1). Figure 4-3 illustrates the coordination roles among the different offices with the DoD and their responsibilities with respect to chemical, biological, radiological, and nuclear (CBRN). Within the DoD, it is the responsibility of the combatant commands (CCMDs) to identify programs and activities of concern by coordinating with the Joint Staff (JS) and Office of the Secretary of Defense (OSD). Joint Force Commanders (JFCs) are tasked with coordinating and cooperating with other USG departments and multinational partners. Communication with the President, NSC, and OSD falls under the responsibility of the Chairman of the Joint Chiefs of Staff (CJCS), which is also designated as the

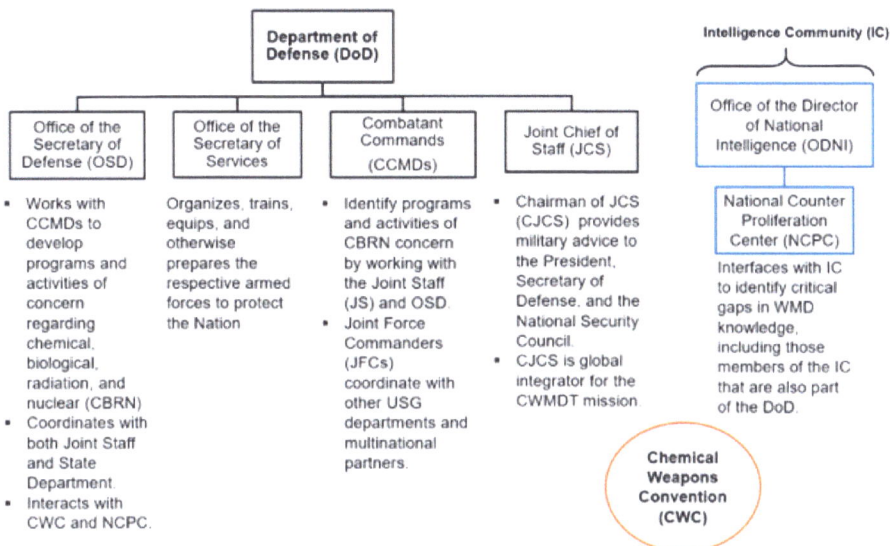

FIGURE 4-3 Organization chart of offices within the DoD and their roles and responsibilities with respect to countering weapons of mass destruction (CWMDT) and chemical terrorism. The DoD's relationship to the IC and the chemical weapons convention is described.

global integrator for the CWMDT mission. The OSD coordinates with both JS and the State Department. Further, OSD interactions with the international Chemical Weapons Convention (CWC), and the National Counter Proliferation and Biosecurity Center (NCBC) are specifically noted as part of their responsibilities.

The NCBC, under the Office of the Director of National Intelligence (ODNI), is charged with coordinating the IC to identify critical gaps in WMD knowledge, including those members of the IC that are also part of the DoD (JCS, 2019). The emphasis within NCBC appears to be in the area of nuclear counter-proliferation.

Since 2018, the United States Special Operations Command (USSOCOM) has led the mission to counter-proliferation of WMD within the DoD. It took over the lead role from U.S. Strategic Command (USSTRATCOM), which retains the lead in strategic deterrence, nuclear operations, and other missions related to nuclear weapons capabilities. USSOCOM is responsible for CCMDs, JS, other DoD agencies, other USG departments and agencies, and partner nations for CWMDT assessment and "transregional synchronization" (JCS, 2019). USSOCOM established the CWMD Fusion Cell, which coordinates planning across organizations (USSOC, 2020). USSOCOM's J10 directorate, based both in the National Capital Region and at USSOCOM Headquarters conducts strategic planning, assesses the department's execution of the CWMD campaign, and makes recommendations to the CJCS and the Secretary of Defense (HASC, 2021). Table 4-1 lists and describes other agencies and programs outside of DoD that have key roles in addressing chemical terrorism.

TABLE 4-1 Key Players Involved in "Identify"

Federal Agency	Programs	Description
National Intelligence Council—Office of the Director of National Intelligence (ODNI)	National Counterterrorism Center (NCTC)	NCTC • integrates and analyzes intelligence pertaining to terrorism including use of WMD • collocates intelligence, military, law enforcement, and homeland security networks to facilitate information-sharing across the USG on terrorist threats • oversees interagency planning for counterterrorism efforts
Department of Justice (DOJ) and Federal Bureau of Investigation (FBI)	Weapons of Mass Destruction Directorate (WMDD)	WMDD provides intelligence support for the FBI field divisions and the rest of the IC on domestic cases. Each field division has a special agent who is the WMD coordinator.
Department of Health and Human Services (HHS)	Agency for Toxic Substances and Disease Registry (ASTDR)	ASTDR works with communities at the local level that are responding to disasters, including those involving hazardous substances

Section 5 of Appendix A of Joint Publication 3-40 explicitly discusses Interagency Coordination and Interorganizational Cooperation (JCS, 2019). For domestic operations, as described earlier, DoD will operate in a supporting role to another USG department or agency. This supporting role is detailed in JP 3-28, Defense Support of Civil Authorities. Interorganizational cooperation is stipulated at the strategic, operational, and tactical levels; the importance of these interactions "cannot be overstated" (JCS, 2019, A-15). When state and local coordination is required, the Chief of the National Guard Bureau (CNGB) will transition to federalized status according to Title 10, U.S. Code,[3] for the CBRN response. Coordination is detailed in JP 3-08, Interorganizational Cooperation, and JP 3-41, Chemical, Biological, Radiological and Nuclear Response. Appendix A of JP 3-40 states that "these processes should be practiced during training events and exercises" (JCS, 2019, A-16).

The discussion above highlights the complexity of information-sharing within and across various agencies involved in identifying chemical terrorism threats. Part of the challenge in the communication network lies in understanding each government agency's roles and responsibilities in CWMDT, which will affect the level and speed of communication. Finding 4-3 underscores the committee's evaluation regarding information-sharing and coordination based on the documents reviewed and briefing presentations.

FINDING 4-3: The agencies surveyed are broadly aware of each other's efforts to identify, prevent, counter, and respond to chemical threats, but express

---

[3] Title 10 of the United States Code specifically pertains to the role and organization of the armed forces, including the Army, Navy, Air Force, Marine Corps, and Coast Guard. It outlines various aspects of military law, regulations, organization, and responsibilities.

concern that information-sharing and coordination across relevant agencies is incomplete.

4.2 "IDENTIFY" STRATEGY EFFICACY

Reviewing historical chemical threats and attacks provides insight into the effectiveness of the IC, LE, and responding organizations in identifying threats. In the United States, there has not been a chemical terrorist event that has had consequences approaching those observed outside of the United States, like the Aum Shinrikyo nerve agent attacks or the Skripal poisonings. (See Appendix E for a description of international case studies.) A review of more recent examples of domestic chemical threats and attacks provides insight into the threat identification challenges for law enforcement. Generally, U.S. response organizations have been effective in identifying chemical threats, in many cases before plans involving the threats have been discovered or the threats themselves have materialized. However, there have been a few notable cases where LE did not identify a threat before an attack was executed, and the 2018 Skripal poisonings in the UK illustrate a new turn in the actors presenting the chemical threat: from that of terrorist-initiated to use by a great power for targeted assassination. A review of selected events (see Appendices F and G for "Threats Interdicted" and "Threats Manifested" case studies) provides a picture of the IC and LE communities' level of success in identifying chemical threats.

The proliferation of information, the ease with which it can be accessed—often anonymously—and the high number of potential perpetrators complicates the task of identifying chemical threats (DSB, 2018). In the case of VEOs, the number of individuals involved makes this task more conspicuous. However, in cases of domestic chemical terror, perpetrators often work alone and do not have a significant footprint either in their communities or online. The availability of information online pertinent to synthesizing and dispersing chemical agents makes exhaustive tracking of all potential perpetrators unlikely. These realities are reflected in the DHS Chemical Defense Strategy, which states: "Detection of chemical threats early in the pathway is very difficult since the pathway steps may be concealed within industrial or agricultural production in dual-use facilities, academic or pharmaceutical labs, dark websites, or private homes or warehouses" (DHS, 2019, Pg. 5). Adding to the complexity of before-the-attack identification is the reality that "industrial chemicals and pesticides are readily available for purchase, and are stored in large quantities in thousands of locations, near population centers" (CRS, 2006, Pg. 10).

Further complicating the identification of chemical threats is the emergence of new chemical agents. These are most likely to emerge from state actors engaging in developing new agents, but the intersection of state and nonstate perpetrators makes the utilization of new agents by VEOs conceivable. DoD is actively engaged in identifying new chemical agents, (DASD(CBD), 2020) but communication of this research (which may be classified) to other agencies may be incomplete.

These realities are overtly stated in the National Strategy for countering WMD, viz., (White House, 2018).

CONCLUSION 4-4a: It is impossible to identify and prevent or counter every threat. Overall, the majority of publicly reported domestic plots did not come to fruition between the 1970s through the mid-2010s for a number of reasons.

CONCLUSION 4-4b: FBI, partner LE, and ICs have been effective in identifying and interdicting the majority of domestic terrorist attacks involving chemical materials, which have typically employed conventional toxic industrial chemicals rather than traditional chemical warfare agents, such as sarin. While the FBI has been effective, approaches to identifying chemical threats could be strengthened using a multilens approach from several different agencies that emphasizes augmented communication and coordination between local and state enforcement and IC. In addition, this area would greatly benefit from increased coordination between the IC and technical experts (particularly those with specific expertise in the areas of terrorist motivation and psychology). For example, FBI antichemical terrorism resources focused on identification could be evaluated in the context of current identification strategies employed by other agencies.

> **RECOMMENDATION 4-4: Existing intelligence community programs should actively seek and incorporate new approaches to identify existing chemical threats (traditional and improvised) and potential emerging threats by terrorist groups. In developing new approaches, program managers should develop strategies that look beyond the traditional terrorism suspects and that augment and leverage skill sets of the USG agencies. For example, scholars of political psychology could work with chemical terrorism experts to create a holistic approach to identifying chemical terrorist groups or similar violent actors outside the traditional suspects. The threat assessments should be improved by reflecting the current times and demographics.**

4.3 IMPLICATION OF THE NATIONAL STRATEGIC SHIFT FROM VEO TO GPC FROM THE PERSPECTIVE OF "IDENTIFY"

The shift in emphasis from threats posed by VEOs to state-sponsored threats arising from GPC (DoD, 2018; Caves and Carus, 2021) further complicates the identification task. For example, it is likely that resources for identifying VEO terrorism threats may be shifted toward GPC threats. An intensification of GPC has many potential implications for chemical terrorist threats beyond a possible reduction of resources available to address them. Even if states simply maintain defensive programs against chemical threats, as many do today, those programs have dual-use implications for offensive threats. Therefore, expertise, technology, and materials, including chemical agents, might be illicitly transferred from defensive programs to nonstate actors, as apparently occurred in the biological weapons domain with the so-called Amerithrax case (DOJ, 2010). States might also choose to engage in offensive chemical warfare agent production—as some states, notably Russia, appear to be doing today—and technology, materials, expertise, and/or chemical agents might be illicitly transferred from those programs to nonstate actors. Further, states, including great powers, might use or support nonstate actors in conducting chemical terrorism. **Witting or unwitting involvement of states in nonstate chemical terrorism could dramatically increase the sophistication of such attacks, including the agent employed or the means by which it is delivered.** Finally, states, including great powers, might engage in the use of chemical agents in ways that might be categorized as "state terrorism"— both Russia and North Korea have done with targeted attacks in recent years—though the committee recognizes that concept is controversial and difficult to clearly define.

FINDING 4-5: The shift to GPC may change the nature of the threat for new chemical attacks, in that chemical agents, other materials, technology, and expertise may migrate from state actors that engage in either defensive or offensive activities to VEOs. These events could enable VEOs to conduct more sophisticated attacks, with agents and/or with means of delivery not otherwise accessible to them.

RECOMMENDATION 4-5: The NCTC, DoD, DHS, and State Department should review current identification approaches to determine whether shifts in emphasis are required as a result of expanded and augmented VEOs and terrorist capability resulting from the potential migration of chemical agents, other materials, technology, and expertise from state actors to VEOs.

One of the biggest risks from the shift to GPC is compromising the USG's ability to adequately identify chemical threats. Attacks stemming from terrorists may look, and may be, substantively and substantially different from an attack caused by a state-based program. Therefore, it is important to make clear distinctions on ways to differentiate state-based and terrorist-based chemical threats within the national strategies.

CONCLUSION 4-6: It is unclear if the tactical readiness to implement the reviewed strategies is occurring at the necessary pace to respond to an act of chemical terrorism. Additionally, the shift in strategic focus to GPC may lead to reduced resources for countering acts of terrorism employing WMDs that are perpetrated by VEOs and may impede tactical readiness against chemical terrorist threats, leading to increased risk.

RECOMMENDATION 4-6: The USG should ensure that the identification of chemical terrorism threats is explicitly included in ongoing and future strategies. Chemical terrorism threats should be considered distinct from nuclear nonproliferation, identification of state-based offensive chemical programs, and traditional (non-nuclear-biological-chemical) terrorism.

4.4 SUMMARY

Federal agencies that briefed the committee acknowledged that the number of potential chemical threats that could be used for WMD is large and increasing. While the FBI, in partnership with LE and the IC, has been effective in identifying and interdicting domestic terrorist attacks involving chemical agents, approaches to identifying these threats could be strengthened. One way to increase capabilities is by using a multilens approach—understanding emerging threats, looking beyond traditional suspects—and increasing coordination and information-sharing between the IC and technical and chemical terrorism experts. Furthermore, the transition toward GPC could alter the risk of new chemical attacks as resources (e.g. chemical agents, technology, and expertise) may transition from state actors to VEOs. As a result, VEOs could potentially carry out more sophisticated attacks using materials and methods that were previously unavailable to them. The committee advised specific programs (NCTC) and departments (DHS, DoD) to review current identification approaches to respond to this potential migration (state actors to VEOs). Because of the shift toward GPC, resources for countering terrorism with WMD may also decrease. Constrained budgetary resources would hinder tactical readiness for implementing the reviewed strategies in response to chemical terrorism. The committee concluded that the USG should include the identification of chemical terrorism threats in ongoing and future strategies, considering them distinct from other nonNBC terrorism threats.

REFERENCES

Casillas, R. P., N. Tewari-Singh, and J. P. Gray. 2023. "Special Issue: Emerging Chemical Terrorism Threats." *Toxicology Mechanisms and Methods* 31(4): 239–241.
Caves, Jr., J. P., and W. S. Carus. 2021. *The Future of Weapons of Mass Destruction: An Update*. National Intelligence University. February, 2021.
Chemical and Biological Defense Program. 2022. Approach for Research, Development, and Acquisition of Medical Countermeasure and Test Products. 2023. https://media.defense.gov/2023/Jan/10/2003142624/-1/-1/0/APPROACH-RDA-MCM-TEST-PRODUCTS.PDF
CRS (Congressional Research Service). 2006. CRS Report for Congress, Chemical Facility Security. August 2, 2006.
DASD (CBD) (Office of the Deputy Assistant Secretary of Defense for Chemical and Biological Defense). 2020. Enterprise Strategy. August, 2020.
Deibel, T. L. 2007. *Foreign Affairs Strategy: Logic for American Statecraft*. New York: Cambridge University Press.
DHS (Department of Homeland Security). 2019. Chemical Defense Strategy. December 20, 2019.
DHS. 2013a. National Infrastructure Protection Plan (NIPP) 2013 Partnering for Critical Infrastructure Security and Resilience.
DHS. 2013b. National Infrastructure Protection Plan (NIPP) Fact Sheet 2013: Partnering for Critical Infrastructure Security and Resilience. https://www.cisa.gov/sites/default/files/2022-11/NIPP-Fact-Sheet-508.pdf
DHS. 2022. The DHS Strategic Plan. Fiscal Years 2020–2024. https://www.dhs.gov/publication/department-homeland-securitys-strategic-plan-fiscal-years-2020-2024.
DoD (Department of Defense). 2014. Department of Defense Strategy for Countering Weapons of Mass Destruction. June 2014.
DoD. 2018. National Defense Strategy of the United States of America: Sharpening the American Military's Competitive Edge.
DoD. 2018. National Strategy for Countering Weapons of Mass Destruction and Terrorism. Presidential Directive, December, 2018. https://www.govinfo.gov/app/details/DCPD-201800841.

DOJ. 2010. "Amerithrax Investigative Summary." https://www.justice.gov/archive/amerithrax/docs/amx-investigative-summary.pdf; Bunn and Sagan (eds) *Insider Threats* (book), chapter on Amerithrax.

DOS (U.S. Department of State). 2022. Compliance with the Convention on the Development, Production, Stockpiling, and Use of Chemical Weapons and on Their Destruction Condition (10)(C) Report. April 2022.

DSB (Department of Defense Science Board). 2018. Task Force on Deterring, Preventing, and Responding to the Threat or Use of Weapons of Mass Destruction. May 2018.

HASC (House Armed Services Committee). 2021. Statement of Vice Admiral Timothy G. Szymanski, U.S. Navy Deputy Commander United States Special Operations Command Before the House Armed Services Committee Subcommittee on Intelligence and Special Operations. May 04, 2021. https://www.congress.gov/117/meeting/house/112537/witnesses/HHRG-117-AS26-Wstate-SzymanskiT-20210504.pdf.

Hersman, R. K. C., S. Claeys, and C. A. Jabbari. 2019. "Rigid Structures, Evolving Threat, Preventing the Proliferation and Use of Chemical Weapons, Center for Strategic and International Studies." December 2019.

JCS (Joint Chiefs of Staff). 2019. Joint Publication 3-40. Joint Countering Weapons of Mass Destruction. November 27, 2019.

NPR (National Public Radio). 2022. "Why Did Tucker Carlson Echo Russian Bioweapons Propaganda on His Top-Rated Show?" https://www.npr.org/transcripts/1089530038.

RAND Corporation. 2007. *Enlisting Madison Avenue: The Marketing Approach to Earning Popular Support in Theaters of Operation*. RAND.

Rotolo, Jason. 2022. FBI, Briefing to the Chem Threats Committee. August 11, 2022. *Prehospital and Disaster Medicine* 34(4) (August): 385–393.

Savage, T. 2022. FBI Policy Program, Briefing to the Chem Threats Committee. August 11, 2022.

Trump, D. 2017. National Security Strategy of the United States of America, National Strategy for Countering Weapons of Mass Destruction Terrorism.

U.S. Joint Forces Command. 2009. *Commander's Handbook for Strategic Communication and Communication Strategy*. https://apps.dtic.mil/sti/pdfs/ADA544861.pdf.

USSOC (United States Special Operations Command). Tip of the Spear. June, 2021. https://www.socom.mil/TipOfTheSpear/USSOCOM%20Tip%20of%20the%20Spear%20June%202021(Web).pdf.

White House. 2018. National Strategy for Countering Weapons of Mass Destruction Terrorism. https://www.hsdl.org/?view&did=819382.

5

Adequacy of Strategies to Prevent and Counter Chemical Terrorism

Summary of Key Findings, Conclusions, and Recommendations

CONCLUSION 5-1: *Upon review of the unclassified strategies, considering United States Government (USG) efforts to dissuade adversaries from pursuing chemical terrorism, there are opportunities to enhance deterrence.*

RECOMMENDATION 5-1: The National Security Council (NSC) should give careful consideration to incorporating direct deterrence of chemical terrorism into existing countering weapons of mass destruction terrorism (CWMDT) strategies.

FINDING 5-2: *There is no evidence of a strategic communications effort that would leverage existing preventive and mitigating measures against chemical terrorism for use as part of a policy of deterrence by denial.*

FINDING 5-3: *Despite ongoing industry practice and some initiatives that previously operated under the Department of Homeland Security's (DHS's) Chemical Facility Anti-Terrorism Standards (CFATS) program, the strategy documents that were made available to the committee do not cite chemical substitution as a key part of an overall chemical security strategy.*

RECOMMENDATION 5-3: Substitution of safer alternative chemicals for hazardous chemicals in industrial and academic settings should be included as part of the overall strategy. The planning and development of these strategies should be spearheaded by DHS's Chemical Information Sharing and Analysis Center under a reauthorized CFATS program and

continued

Summary Continued

should continue to be conducted in conjunction with regulatory agencies, specifically, the Environment Protection Agency (EPA), and the Occupational Safety and Health Administration (OSHA), as well as representatives from industry and academic research environments.

FINDING 5-4a: The strategic documents surveyed do not explicitly mention insider threat in the chemical terrorism context.

FINDING 5-4b: While CFATS included some practical efforts to counter insider threats within the chemical industry, the scope of these efforts appeared to be limited and the committee did not find evidence of a similar systematic program either directed towards government facilities or within academic research institutions.

CONCLUSION 5-4: The significant potential consequences of an insider at a chemical facility conducting or assisting an attack warrants explicit inclusion in existing strategies and comprehensive policies to counter insider threats at any facility containing significant quantities of toxic chemicals.

RECOMMENDATION 5-4: Counter-insider-threat activities should be incorporated explicitly into broader counter weapons of mass destruction (WMD) strategy. The Office of the Director of National Intelligence (ODNI), with other public and private partners, should develop a strategy to ameliorate insider threats explicitly for the chemical domain.

FINDING 5-5: The committee found that certain key activities that the USG is appropriately undertaking were surprisingly absent from the strategy documents reviewed, including: military capabilities to provide early warning of chemical terrorism plots; law enforcement (LE) capabilities to counter chemical threats tactically; integration with broader counterterrorism and countersmuggling efforts; and involvement with other multilateral activities beyond the Organization for the Prohibition of Chemical Weapons (OPCW).

CONCLUSION 5-5: The committee concludes certain key activities that are undertaken in practice to prevent and counter chemical terrorism are sufficiently important to merit inclusion in strategy documents. The absence of such activities from the strategies could impact policy implementation, including budgeting, program prioritization, and other consequences. Including these activities in existing strategies would bolster the comprehensiveness, and therefore effectiveness, of existing strategies.

RECOMMENDATION 5-5: Agencies should work to reconcile operational practice with policy by supplementing extant strategies to include current omitted effective activities and programs for countering chemical terrorism. This would ensure that effective practices are maintained, properly resourced, and reflected in comprehensive strategies.

The committee surveyed the following strategy documents listed in Box 5-1. All of these documents contained useful information related to aspects of preventing and countering chemical terrorism.

5.1 ANALYSIS OF STRATEGIES TO "PREVENT OR COUNTER" CHEMICAL TERRORISM THREATS

Most of the strategy documents espoused a coherent strategy or set of strategy elements comprising a combination of a well-defined goal with a corresponding definition of success, as well as at least one policy, plan, and/or resource allocation designed to meet the goal. The exception to this assessment was a DoD Directive 2060.02, which did not provide clear definitions of success for its goals of "dissuade, deter, and defeat actors' concern and their network; [. . .] manage WMD terrorism risks from hostile, fragile or failed states and safe havens; [or] limit the availability of WMD-related capabilities." (Pg. 3)

There have also been more recent strategic documents that touch on chemical terrorism not available to the committee. The Executive Office of the President (EOP) issued classified documents describing the USG's internal organization and policies on actions to prevent, counter, and respond to WMD terrorism. The most recent is the *National Security Memorandum (NSM) 19 to Counter Weapons of Mass Destruction (WMD) Terrorism and Advance Nuclear and Radioactive Material Security.* Although NSM-19 is classified, government officials have released some information about it

---

BOX 5-1
Strategy Documents Reviewed by Committee
for "Prevent/Counter" Analysis

1. White House. (2018). National Strategy for Countering Weapons of Mass Destruction Terrorism.
2. Department of Homeland Security. (2019). Department of Homeland Chemical Defense Strategy.
3. Department of Defense Directive 2060.02. (2017). DoD Countering Weapons of Mass Destruction (WMD) Policy.
4. Department of Defense. (2014). Strategy for Countering Weapons of Mass Destruction.
5. Joint-Publication 3-40. (2021). Joint Countering Weapons of Mass Destruction.

NOTE: At the time of writing this report, the DoD released the *2023 Department of Defense Strategy for Countering Weapons of Mass Destruction*; the committee did not evaluate this document.

(The White House, 2023) and have given an unclassified briefing on its contents, which are meant to address what the administration sees as the changing features of the terrorist threat. Elizabeth Sherwood-Randall issued a statement at a Nuclear Threat Initiative event on March 2, 2023: "It [the terrorist threat] has become more ideologically diffuse and geographically diverse." Much of the presentation focused on nuclear terrorism; however, chemical terrorism was referenced: "threats that are not existential, but more likely, such as chemical and radiological terrorism" (Sherwood-Randall, 2023).

As explained in Section 5.1.1, the committee evaluates the adequacy of strategies by whether they contain certain elements and features. Most of the strategy elements that the committee believes are essential to address both the current and emerging threat of chemical terrorism are reflected in the above-mentioned strategy documents. The method described in Chapter 3 lists these elements and whether each one is addressed by the strategies.

5.1.1 Committee's Definition of Adequacy: Prevent or Counter

The committee argues that a successful strategy to prevent or counter chemical terrorism focuses on the following elements:

- Incorporates developments in the "Identify" area into practice for "Prevent and Counter."
- Dissuades terrorists through deterrence by denial, deterrence by punishment, or through normative means.
- Impedes acquisition of raw materials, production technology, delivery technology, or information for production or delivery. Strategy also demonstrates having mechanisms (e.g., insider threat programs, strategic trade controls, international efforts, and collaboration with other counterterrorism programs) to ensure that those items are not acquired.
- Interdicts active plots through military, LE, or intelligence capabilities.
- Ensures collaboration at various levels—international, federal, state, local, tribal, and territorial (SLTT).
- Addresses new chemical terrorism threats; new chemical agents, new production or delivery methods and technologies, and new actors.
- Forms collaboration with nonterrorist focused agencies (e.g., Drug Enforcement Agency [DEA]).

The committee's analysis considered all six of these major elements. Successful "prevent/counter" strategies focus on communicating clearly which adversaries they will prevent and counter from committing acts of chemical terrorism (often described in the documents' "goals" and "objectives") and lay a clear plan for how the respective agencies will ensure that their goals are met.

5.1.2 Countering Identified Threats

The Federal Bureau of Investigation (FBI) looks at all actors (state, nonstate, and lone), all modalities, and all stages from pre-event to post-event. The Weapons of Mass Destruction Directorate (WMDD) has programs for preparedness, countermeasures, investigations and operations, and intelligence (coordinated with the FBI Intelligence Branch) (Savage, 2022).

The WMD Intelligence Analysis Section focuses on proliferation threats from state actors and counterterrorism activity concerning nonstate actors such as al-Qa'ida, al Shabab, and others. The chemical, biological, radiological, and nuclear (CBRN) Intelligence Unit in the FBI's Intelligence Branch includes technical experts who focus on threats and vulnerabilities of the materials themselves. They will investigate "any actor acquiring or seeking to acquire chemical, chemical expertise, related or emerging technologies for use, threatened use, or attempted use as a weapon" (McNelis, 2022).

The Chemical Biological Countermeasures Unit (CBCU) works to prevent (detect, deter, and disrupt) WMD attacks, with a major focus on outreach. They conduct outreach through the Chemical Industry Outreach Workshop,[1] Livewire Exercises,[2] the Chemical Facility Outreach Program,[3] WMD coordinators in field offices, and partnerships with other government agencies such as former CFATS and Flashpoint,[4] implemented by the DHS's Cybersecurity and Infrastructure Security Agency (CISA). There are 56 WMD coordinators, one in each of the 56 FBI field offices, and there are FBI WMD specialists present in 90 countries. WMD coordinators are FBI special agents who work with partners in other LE organizations and the private sector. They are trained in chemical, biological, radiological, nuclear, and high yield explosives (CBRNE) investigations and promote two-way information-sharing. They are supported by 300 assistant coordinators. The CBCU also works to protect advances in scientific research and technology development from theft and misuse (Sharp, 2022).

The FBI has a biannual threat review process where they rank threats, and a program within the bureau tracks how well each field office is doing to mitigate threats based on their prioritization. This helps in evaluating effectiveness and prioritizing among the various programs. The FBI WMDD has within its scope chemical (warfare agents to industrial and household chemicals), biological (pathogens and toxins), radiological, nuclear, and explosive threats (CBRNE). As a result, in addition to addressing the traditionally understood WMD threats, the WMDD CBCU works to prevent and protect against the use of explosives, with the intention to raise awareness with vendors

---

[1] A workshop to educate LE, first responders, chemical manufacturers, retailers, distributors, and academia regarding explosive precursor chemical (EPC) products that may be used to manufacture explosives.

[2] Livewire is a tabletop exercise of response to a chemical terrorist attack. People from all organizations involved in the response come to the table to exercise a response.

[3] A regional workshop to engage chemical facilities that manufacture, store, use, transport, or distribute chemicals of interest on threat-related issues.

[4] Flashpoint's goal is to raise awareness within stores selling precursor chemicals.

selling explosive precursor chemicals (EPCs) and providing information on "who to contact" if suspicious purchases are occurring.

Capabilities for countering CWMD within the FBI require significant intellectual resources. Attracting personnel with the ability to address the technological issues derived from a chemical terrorism event is a challenge for the FBI (Savage, 2022). They are able to mitigate this problem to some extent by contracting needed expertise from organizations outside the FBI, and by forming partnerships with sister organizations, but there are areas where more personnel are needed. One area in particular need of augmented personnel is data science. The FBI has had the same staffing level for the past 17 years, and in some areas, like data science, would benefit from increased staffing levels.

5.1.3 Deterrence or Reducing Motivation

Upon reviewing existing strategy documents, the committee found references to deterrence by punishment in a nonspecific context. For example, the 2002 National Strategy to Combat Weapons of Mass Destruction (3) reiterates the declaratory policy that:

> The United States will continue to make clear that it reserves the right to respond with overwhelming force—including through resort to all of our options—to the use of WMD against the United States, our forces abroad, and friends and allies[. . .] posing the prospect of an overwhelming response to any use of such weapons.

The overall document explicitly cites terrorists as a source of potential risk in the context of the acquisition and use of WMD, but they are not explicitly called out in the context of deterrence.

Strategies addressing nonstate actors appear to be focused predominantly on other forms of deterrence, which could involve threatening to punish potential states, nonstate institutions, and even individuals who might support terrorists acquiring WMD (including chemical weapons). The committee found no explicit declaratory statement of direct deterrence by punishment directed toward terrorists who used chemical weapons, in contrast to both the nuclear and biological domains.

Deterrence of chemical terrorism is also different in that the United States has, in the past, responded to actionable intelligence indicating a terrorist group was attempting to obtain chemical weapons with preemptive strikes (Croddy, 2002).

As discussed in Box 5-2, direct deterrence of terrorists, while not straightforward, is possible. Moreover, there are substantial advantages to an explicit communication of the direct deterrence proposition (e.g., that the United States will take certain measures if terrorists utilize chemical weapons that would not otherwise be taken). This is because, in order for deterrence to work at all, the deterrence proposition needs to be conveyed effectively to the targets of the deterrence. For states, this can be done publicly through pronouncements, privately through diplomatic and other channels, or implicitly through exhibited capabilities or actions. However, in the case of terrorists, which represent a diffuse set of actors scattered around the world with whom the United

States rarely has formal (or even informal) relations, the efficacy of private channels is likely to be low at best.[5]

While the proposition can be conveyed implicitly and generally (e.g., if the U.S. military has conducted expanded drone strikes within terrorist safe havens in response to a previous escalation in the scale or scope of terrorist attacks), this approach might not be sufficient to deter a chemical attack specifically. This is because deterrence by punishment relies in most cases on reducing the adversary's ambiguity about the severity, celerity, and certainty of the reprisal and thus enhancing credibility. Although most terrorists would infer based on past U.S. behavior that the United States would retaliate vigorously to a large-scale chemical attack, they might not realize how vigorously or how quickly it would react.[6]

One possible reason for not explicitly communicating a direct deterrence proposition (e.g., drawing a red line) is that it can tie the hands of the USG because once a threat is made, it must be followed through on, or further deterrence is undermined. It is arguable that the failure of the Obama Administration to respond forcefully when Syria crossed the President's stated "red line" by using chemical weapons against its people undermined U.S. chemical weapons deterrence vis-à-vis states, and policymakers may be reluctant to repeat this episode in the terrorism context.

There are classified EOP policy and guidance documents relating to deterrence of chemical weapons use worldwide, including both state and nonstate threats. When the committee asked why public versions of these policies to address chemical terrorism threats have not been issued (given that there are such documents for nuclear and biological threats) and that existing strategies do not mention direct deterrence of terrorists, the EOP replied that the classified documents provide what the White House needs from them—mostly direction for internal policy and organization. The nuclear and biological threats are addressed in separate major strategy documents, the nuclear posture review and the biodefense strategy, for example. No such document is required by law for chemical threats. Further explanations were not provided to the committee, nor is there any evidence that direct deterrence (at least explicitly conveyed) is a major part of current strategies, either in the classified or unclassified realm. That said, officials informed the committee that the United States' position concerning chemical weapons and chemical terrorism is conveyed through other means such as the Chemical Weapons Convention (CWC) compliance reports, support for the Organization for the Prohibition of Chemical Weapons (OPCW) and the 1540 committee,[7] and other actions. The implication appears to be that through multiple means the USG conveys that it seeks to, and demonstrates that it does, hold responsible for its actions anyone who commits

---

[5] A declaratory policy does not have to be enshrined in an official executive order or national strategy. A statement in a speech by the President or the Secretary of Defense might be enough.

[6] Ambiguity about what scale of chemical use would trigger punishment is valuable. Establishing clear thresholds or red lines leads adversaries to approach those thresholds without exceeding them.

[7] The 1540 committee monitors and supports the efforts to implement UN Security Council Resolution 1540, passed in 2004. The resolution requires all nations to institute domestic legal-regulatory measures and controls to prevent the proliferation of nuclear, chemical, and biological weapons of mass destruction, their means of delivery, and related materials to non-State actors (DOS, 2022).

BOX 5-2
Background on Deterrence

In the immediate aftermath of the September 11, 2001, terrorist attacks, it was often suggested—including by President George W. Bush and senior members of his administration—that, unlike states, terrorists were undeterrable, both because many had no fixed addresses and because many were suicidal. There was subsequently considerable scholarly and analytic pushback against this notion, and mainstream assessments today posit that deterrence is an important part of the counterterrorism toolkit, and therefore also the counterchemical terrorism toolkit.

Deterrence is an influence strategy, trying to dissuade the other side from undertaking some action through the use of negative incentives. It most commonly refers to the use of conditional threats, where the costs threatened are intended to outweigh the benefits from the action being considered, labeled "deterrence by punishment." More broadly, "deterrence by denial" involves denying the attainment of benefits so that the actor is dissuaded from attempting the action in the first place. Some policies may fall into both categories; for example, security guards in stores both make it more likely shoplifters will be foiled in the attempt and also that they will be punished. On the other hand, some policies fit more neatly into one category; for example, signs in stores indicating that "all shoplifters will be prosecuted" (despite the cost to the store) are a pure deterrence by punishment threat. Similarly, the actions of both targeting perpetrators and massive retaliation (reprisal using disproportionate force) after an attack demonstrate punishment, and beyond simple revenge, presumably these are intended to deter future attacks.

In order for terrorists and their supporters to be deterrable, several conditions must apply, and all of them are so-called "necessary conditions," meaning they must all be met. To be deterrable:

1. Terrorists or supporters must have preferences, such as having things they care about, and to varying degrees must care more about some things than other things. This is labeled either "rationality" or "minimal rationality."
2. Those seeking to deter terrorists must be capable of affecting those preferences, which implies knowing what those preferences are.
3. Threats of denial and punishment must be credible.
4. Effective implicit and/or explicit communication must occur, such that the target both receives and understands the threat (in the case of deterrence by punishment) or the increased likelihood

of not achieving their objectives (in the case of deterrence by denial).

Or the government may genuinely have few options for escalation: if the government already has maximal campaigns against certain groups or individuals, then they cannot be increased. And no approach to deterrence works in all cases. In practice, presumably some terrorists are not deterrable, but many are, and their supporters are even more so.

While there have been lively debates about whether terrorists are deterrable, there is a widespread consensus that potential state sponsors are deterrable. Relatedly, there have been debates about what some have termed "deterrence of negligence," with a focus on nuclear terrorism threats, though the same arguments are potentially applicable to chemical terrorism. The argument is that if states are lax in controlling materials or weapons, leading them to fall into the hands of terrorists, those states should be held accountable for the consequences of their actions. While states might perceive implicit versions of the threat, the argument is sometimes made that the threat would be more effective if it were articulated explicitly, perhaps publicly, perhaps via private channels.

Criminal justice literature suggests that deterrence by punishment effects vary based on the likelihood, severity, and celerity (e.g., speed of punishment). It is intuitive that the likelihood of facing punishment, and the severity of punishment is important, and less intuitive than celerity (i.e., the time it takes for the punishment to be imposed) should also be important.

The implicit and explicit threats that undergird deterrence by punishment may backfire under some circumstances. Some actors may be less concerned about, or may even welcome, harsh responses. For example, some argue that Al Qaeda welcomed the U.S. invasion of Afghanistan. Far-right extremists might welcome a harsh crackdown by government authorities, thinking it will generate sympathy and support for their cause or spark an uprising.

Separately, albeit relatedly, some actors may also welcome the opportunity to send so-called "costly signals" to various audiences, and making them more costly may therefore induce rather than deter action. For example, terrorists might assess that if they are seen as boldly taking on the powerful USG, despite the backlash they are risking, that will bolster their recruitment efforts. Threatening a more intense backlash might only intensify these motivations. Alternately, terrorists might calculate that if they are seen as successfully attacking a particularly well-defended target it will redound to their benefit.

a terrorist attack against the United States, including chemical attacks. The Committee argues, however, that U.S. statements through international bodies such as the OPCW or the 1540 Committee, while potentially useful in indicating U.S. positions to other states, are unlikely to constitute sufficient explicit conveyance of a deterrence proposition regarding chemical attack to terrorist actors.

The committee judges that careful consideration should be given to incorporating direct deterrence of chemical terrorism into existing CWMDT strategies. More assessment may be needed to determine: a) whether the direct deterrence by punishment strategy needs to be communicated explicitly for chemical terrorism, or b) whether implicit, more generalized threats that can be inferred by terrorists might suffice to accomplish most of the deterrence that can be achieved. Such a study would also include discussions on the strategic and operational implications of issuing an official document versus including a declaration in a national security speech or via another channel.

CONCLUSION 5-1: Upon reviewing the unclassified strategies, considering USG efforts to dissuade adversaries from pursuing chemical terrorism, there are opportunities to enhance deterrence.

RECOMMENDATION 5-1: The NSC should give careful consideration to incorporating direct deterrence of chemical terrorism into existing CWMDT strategies.

5.1.3a Deterrence by Denial

The committee found multiple existing policies and programs that contribute to a strategy of deterrence by denial. These include facility security improvements under CFATS[8] and a variety of response capabilities (see Chapter 6) that would mitigate the harm caused by a chemical attack. For maximum deterrence effects to be achieved, it is therefore valuable to craft and implement a specific communications strategy conveying the efficacy of defensive capabilities[9] to convince would-be adversaries that they will not achieve the goals they seek from using chemicals. The amount, content, and channels associated with such a communications effort require careful consideration and crafting. On the one hand, leveraging deterrence by denial for chemical attacks depends on informing terrorists why their plots are unlikely to succeed, which means that some information about defensive measures must be released if we want to dissuade them from trying in the first place. On the other hand, providing too much detail about defenses and response capabilities could potentially aid terrorists in circumventing those same defenses. More study is needed beyond the scope of the committee.

Yet, irrespective of the specific nature of any communications meant to enhance deterrence by denial, the committee could find no evidence of a dedicated communica-

---

[8] At the time of writing this report, the statutory authority for the CFATS program (6 CFR Part 27) has expired and has yet to be reauthorized.

[9] There is the possibility of intentionally mischaracterizing this effectiveness, for example exaggerating the ability to detect or treat chemical use to enhance the deterrent effect, but this carries risks of undermining deterrence when the mischaracterization is discovered.

tions strategy or policies to this effect in any of the strategic documents or other materials it reviewed. The committee understands that it is inadvisable to disclose all security measures that exist to prevent or mitigate chemical terrorism. Nonetheless, since the United States is already engaging in these activities, and some of them are already known publicly (albeit in relatively arcane circles) the United States could leverage its achievements in these areas as part of a deliberate strategic communications effort to enhance deterrence by denial. For example, there is no explicit reference in public strategy documents to improving the robustness or lowering the vulnerability of civilian populations and targets in general to chemical attack, even though there are numerous programs that are doing just this. Such a strategic communications effort could be coordinated and implemented by CISA domestically and the State Department's Global Engagement Center and Bureau of International Security and Nonproliferation abroad, working with the NSC.

> FINDING 5-2: There is no evidence of a strategic communications effort that would leverage existing preventive and mitigating measures against chemical terrorism for use as part of a policy of deterrence by denial.

5.1.4 Reducing Material Availability

Some chemicals are readily accessible, others far less so, with a spectrum from extremely accessible (e.g., commercially available household chemicals), relatively accessible (e.g., many so-called toxic industrial chemicals (TICs) present in chemical plants and manufacturing facilities), to extremely inaccessible (e.g., organophosphate nerve agents and many of their key precursor chemicals). In theory, any chemical can be produced from readily available precursor chemicals. In practice, the technical barriers to producing certain chemicals are high, in some cases extremely high. Many policy practitioners and the general public seem not to appreciate the notable challenges associated with organophosphate nerve agent synthesis, even if key precursor chemicals are obtained, for example (Kaszeta, 2018).

Accessibility is shaped to a large extent by the domestic legal and regulatory structure, which is in turn influenced by international law and efforts at export control harmonization via multilateral export control regimes, as well as private sector controls independent of/in addition to legal and regulatory requirements. The United States also helps other countries bolster their controls, as part of a broader international architecture put in place to try to identify and ameliorate the weakest links in an international chain of chemical supply and production that potentially can put everyone at risk.

5.1.4a Regulatory efforts to reduce material availability (domestic & international)

Domestically, Executive Order 13650 on Improving Chemical Facility Safety and Security, issued on August 1, 2013 (Exec. Order, 2013) directed the federal government to:

- improve operational coordination with state and local partners; enhance Federal agency coordination and information-sharing; modernize policies, regulations, and standards; and work with stakeholders to identify best practices.

The order established the Chemical Facility Safety and Security Working group, co-led by DHS, OSHA, and EPA working with the Bureau of Alcohol, Tobacco, Firearms and Explosives (ATF), the Department of Transportation (DOT), the Department of Agriculture (USDA), SLTT governments, first responders, industry, and community stakeholders.

Table 5-1 lists the federal agencies and their respective programs that are involved in countering chemical terrorism. Under this executive order, U.S. EPA established the Risk Management Program regulation, which aims to "reduce the likelihood of accidental releases at chemical facilities, and to improve emergency response activities when those releases occur" (Exec. Order, 2018). DHS's CISA established CFATS. Chemical substitution programs at EPA and former CFATS programs are further discussed in section 5.1.4b "Chemical Substitution." Additional information regarding CFAT's authorization expiration can be found in section 7.5.

Internationally, the OPCW is the implementing body for the CWC. The CWC "restricts the production of many 'dual-use' chemicals, such as those that could be used both in the illegal production of chemical weapons and for peaceful chemical processes." (OPCW, n.d.)[10] To prevent dual-use chemicals from being misused, States Parties to the CWC have committed to ensure that "all toxic chemicals, and their precursors, are only used for purposes that are not prohibited by the Convention." (OPCW, n.d.). The signatory declares quantities and types of chemicals and submits the declarations regarding these chemicals. OPCW inspects "the facilities where these chemicals are produced, processed, or consumed to ensure that the declarations are complete and accurate." (OPCW, n.d). Also, when they export or import scheduled chemicals, States Parties to the Convention are obliged to declare international transfers between States Parties and are prohibited from trade in certain chemicals with nonstates Parties (OPCW, n.d.).

The CWC also requires its States Parties to put in place "controls, where considered necessary, on scheduled or non-scheduled chemicals that are susceptible to being used as weapons or in the manufacture of chemical weapons" (OPCW, n.d.). This includes physical protections and regulatory requirements.

In 2004, the United Nations Security Council adopted resolution 1540 (UNSCR 1540) (UNSC, 2004) which obliges states to "refrain from providing any form of support to nonstate actors that attempt to develop, acquire, manufacture, possess, transport, transfer or use nuclear, chemical or biological weapons and their means of delivery." It goes on to declare that states need to adopt and enforce laws prohibiting the WMD-related actions listed above, establish appropriate controls over related materials, including accounting for and securing items and materials, and develop and maintain appropriate effective border controls and efforts to detect, deter, prevent and combat the

___

[10] There is a nuance here. If the chemicals are used for peaceful/allowed purposes under the CWC, the production is not so much restricted as it is monitored and verified to be used for the declared purpose. The CWC allow chemical warfare agents to be produced and possessed when they are being used for research purposes to counter chemical agents (e.g., analysis methods, personal protective equipment (PPE), and countermeasure development and testing, etc.). Additionally, CWC Schedule III chemicals be produced in large quantities.

TABLE 5-1 Key Players and Roles in "Prevent/Counter"

Federal Agency	Programs	Description
Department of Homeland Security (DHS)	• Office of Infrastructure Protection (OIP) • Office of Intelligence and Analysis (I&A) • Office of Bombing Prevention (OBP) • Chemical Facility Anti-Terrorism Standards (CFATS)*	• OIP oversees critical infrastructure protection, including chemical facilities. • I&A provides intelligence support and threat assessments related to chemical terrorism. • OBP focuses on preventing and responding to explosive threats involving chemical components. • CFATS ensured the security of high-risk chemical facilities by implementing risk-based performance standards and regulations.
Department of Justice (DOJ) and Federal Bureau of Investigation (FBI)	• Weapons of Mass Destruction Directorate (WMDD) • Bureau of Alcohol, Tobacco, Firearms and Explosives (ATF)	• FBI investigates and disrupts chemical terrorism threats and attacks, working in coordination with other agencies. More specifically, WMDD investigates and disrupts chemical terrorism threats, including the use of chemical weapons or agents. • ATF addresses the illegal use, acquisition, and trafficking of chemicals, including those that can be used for terrorism.
Environmental Protection Agency (EPA)	• Chemical Emergency Preparedness and Prevention Office (CEPPO) • Criminal Investigation Division	• CEPPO implements the Risk Management Program (RMP) to prevent and prepare for chemical accidents and incidents with potential terrorism implications. • The Criminal Investigation Division investigates violations of environmental laws related to chemicals, including those with potential terrorism connections.
Department of Defense (DoD)	• Defense Threat Reduction Agency (DTRA • Chemical and Biological Defense Program (CBDP)	• DTRA develops and deploys advanced detection and response capabilities to counter chemical and biological threats, including terrorism. • CBDP works to protect military personnel and the nation against chemical threats, including those from terrorist activities.
Department of State (State)	• International Security and Nonproliferation Bureau (ISN) • Office of Cooperative Threat Reduction (CTR) • Export Control and Border Security Program (EXBS) • Office of Counterterrorism (CT) • Office of Export Control Cooperation (ECC) • Office of Multilateral Nuclear and Security Affairs (MNSA) • Office of the Nonproliferation and Disarmament Fund (NDF) • Office of Weapons of Mass Destruction Terrorism (WMDT)	• State is the USG's primary interface for interactions with other nations and with international organizations. In addition to leading diplomacy and having a role in policy, State has several implementation programs for everything from identification and prevention through consequence management under several offices, including CTR, ECC, CT, and others.

continued

TABLE 5-1 Continued

Federal Agency	Programs	Description
Department of Health and Human Services (HHS)	• Office of the Assistant Secretary for Preparedness and Response (ASPR) • Centers for Disease Control and Prevention (CDC)	• ASPR coordinates medical and public health emergency preparedness and response efforts, including those related to chemical terrorism. • CDC conducts research, surveillance, and response activities related to chemical threats and public health impacts.
Department of Energy (DOE)	• National Nuclear Security Administration (NNSA)	• NNSA works to prevent, detect, and respond to radiological and nuclear threats, including those involving chemical agents.
Department of Agriculture (USDA)	• Animal and Plant Health Inspection Service (APHIS)	• APHIS protects agricultural and natural resources from potential chemical threats, including agroterrorism.
Department of Transportation (DOT)	• Pipeline and Hazardous Materials Safety Administration (PHMSA)	• PHMSA regulates the safe transportation of hazardous materials, including chemicals, to prevent incidents and potential terrorist exploitation.
Food and Drug Administration (FDA)	• Center for Drug Evaluation and Research (CDER)	• CDER ensures the safety and security of pharmaceuticals, including controlled substances used in the treatment of opioid addiction.

NOTE: At the time of writing this report, CFATS statutory authority expired.

illicit trafficking of materials and items related to WMD. To help implement UNSCR 1540, the resolution established the 1540 Committee.

The Proliferation Security Initiative (PSI), launched by the United States in 2003, is a cooperative international effort "to stop trafficking of weapons of mass destruction, their delivery systems, and related materials to and from states and nonstate actors of proliferation concern." One hundred and six countries have signed on to PSI and its Statement of Interdiction Principles, committing to "interdict transfers to and from states and nonstate actors of proliferation concern to the extent of their capabilities and legal authorities." They also commit to exchange information and "take specific actions in support of interdiction efforts" (DOS, 2003).

The chemical weapons-relevant CTR are situated in the DTRA and the State Department's ISN (Becker and Nalabandian, 2022). The original objective of CTR was to work with former Soviet states to destroy weapons of mass destruction, including chemical weapons. The scope of CTR has expanded to include work with other cooperative states to reduce the chemical threat, including both terrorism and state-level activities.[11] Most of DTRA's CTR activities are focused on the elimination of known

---

[11] See 2009 Global Security Engagement: A New Model for Cooperative Threat Reduction; 2018 Cooperative Threat Reduction Programs for the Next Ten Years and Beyond Proceedings of a Symposium—in Brief; 2020 A Strategic Vision for Biological Threat Reduction: The U.S. Department of Defense and Beyond.

stockpiles of chemical warfare agents. This reduces the probability of a terrorist chemi-cal terrorism event by reducing the overseas availability of chemical warfare agents and their precursor chemicals. Other reductions are made through know-your-customer pro-grams for businesses, and a variety of other activities and training or capacity-building programs that parallel domestic programs. Because of the possibility that precursor chemicals could be transported to a VEO, CTR programs also are taking a more active interest in chemical transportation security.

As stated in the name, the programs are cooperative, working with other nations and international organizations—International Criminal Police Organization (INTERPOL), United Nations Office on Drugs and Crime (UNODC), and others—on shared priori-ties. In some cases, noncooperative regimes change (e.g., Libya), creating significant risks and opportunities. Together with the Nonproliferation and Disarmament Fund (described in Chapter 7), CTR programs can secure and eliminate stockpiles before they fall into the hands of VEOs. The network of experts in, and associated with, the CTR program and their partners is a resource to exploit opportunities and mitigate risks quickly when they arise. At the same time, it is difficult for under-resourced partner countries to sustain the capabilities that CTR has provided. Maintaining knowledgeable staff, both in the United States and in partner countries, is also a challenge.

5.1.4b Chemical Substitution

Another key avenue by which the risk of chemical terrorism threats can be reduced is to replace existing processes and materials with less toxic alternatives, often referred to as inherently safer technology (IST). This chemical substitution reduces the potential consequences of a chemical terrorist attack by making toxic materials less prevalent or by eliminating their use entirely, making theft and use of the materials in commerce more difficult and less attractive, and making the facilities less attractive targets for sabotage (see above discussion of deterrence by denial). While terrorism threat would be lowered, a large-scale chemical release will still cause disruption even though the toxicity is low (e.g., time for cleanup, environmental impact, and other disruptions).

Chemical substitution and IST are not a new idea. Occupational and environmental safety concerns have long driven industry to seek substitution as a strategy to mitigate hazards, and both the OSHA and EPA have for decades encouraged and recognized innovative approaches for substitution,[12] with a number of programs and offices play-ing a role.[13] Indeed, OSHA provides a graphic on its website (see Figure 5-1 below) emphasizing that reduction/elimination of hazardous substances is the most effective risk-reduction mechanism (OSHA, n.d.).

Similarly, the EPA has, as part of its Risk Management Program (RMP), for a long time collaborated with industry and academia to encourage substitution as part of a

---

[12] On its website https://www.osha.gov/safer-chemicals, OSHA notes that "Where possible, elimination or substitution is the most desirable followed by engineering controls".

[13] Examples include EPA's Chemical Safety and Sustainability Research Program, the Center for Public Health and Environmental Assessment, and the National Institute of Environmental Health Sciences.

transition to "green chemistry."[14] For example, since 1996, the EPA's Office of Chemical Safety and Pollution Prevention has sponsored the Green Chemistry Challenge Awards in partnership with the American Chemical Society's Green Chemistry Institute® and other members of the chemical industry.[15] Yet, safety and security—while overlapping to some degree—have different aims, and intentional harm can manifest in very different ways from accidental or environmental harm. Neither the EPA nor the OSHA programs are focused on security. For example, they do not include theft or diversion as considerations for decisions regarding substitution. Other security risks involving chemical substitution also include the possibility of the *less* harmful chemicals being used as precursors to synthesize *more* toxic chemicals.

DHS's Chemical Facility Anti-Terrorism Standards (CFATS) program did implicitly encourage chemical substitution and lowering inventories of toxic chemicals since facilities with less toxic chemicals have lower reporting and security requirements. Even though the chemical industry has vigorously opposed any mandatory IST as part of the CFATS program,[16] a productive partnership was developed between CFATS and industry regarding voluntary chemical substitution efforts, with substantial reductions in the usage of hazardous chemicals (Subcommittee on Cybersecurity and Infrastructure Protection, 2018). These partnerships likely translated into an overall reduction in risk. Yet, these laudable efforts by CFATS and industry are not embodied in any larger government strategy—none of the strategy documents reviewed by the Committee mention IST or chemical substitution as a distinct goal or strategy.

Absent a coherent strategy, innovations that might enhance or facilitate existing practice may not be resourced or encouraged. For example, making substitution an explicit part of the government's strategy could allow for such measures as direct incentive policies for industry to substitute less-toxic chemicals or processes, as well as encouraging research into which substitutions would have the greatest security, in addition to safety and environmental, benefits. As a comparable case study in a related domain, replacing certain dangerous radioisotopes in industry, medicine and research is a strategic goal of the DOE/NNSA, embodied in the Cesium Irradiator Replacement

---

[14] Green chemistry is defined by the EPA as: "the design of chemical products and processes that reduce or eliminate the use or generation of hazardous substances. Green chemistry applies across the life cycle of a chemical product, including its design, manufacture, use, and ultimate disposal." It further states that "Green chemistry reduces pollution at its source by minimizing or eliminating the hazards of chemical feedstock's, reagents, solvents, and products." [https://www.epa.gov/greenchemistry/basics-green-chemistry]

[15] The EPA website notes that "through 2022, our 133 winning technologies have made billions of pounds of progress, including [. . .] 830 million pounds of hazardous chemicals and solvents eliminated each year—enough to fill almost 3,800 railroad tank cars or a train nearly 47 miles long" [https://www.epa.gov/greenchemistry/information-about-green-chemistry-challenge.

[16] When IST requirements were included in the House version of the 2009 CFATS bill, both the American Chemistry Council and the Society of Chemical Manufacturers and Affiliates came out in opposition, arguing that the concept of IST was nebulous and that DHS lacked the capacity to understand all of the downstream effects of a particular mandate on industry supply chains, which could lead to disruption and shortages of important products. Mandatory IST was not included in the final statute. See: https://www.pharmtech.com/view/house-committee-passes-ist-requirements-chemical-facility-security-bill.

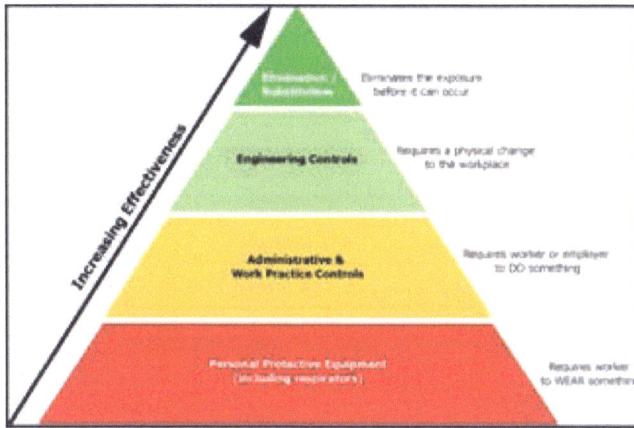

FIGURE 5.1 OSHA Chemical Risk Reduction Pyramid (OSHA, n.d).

Project (Lieberman and Itamura, n.d.). While this project covers only a limited number of hazardous substances, and there are far fewer facilities housing dangerous radioisotopes than dangerous chemicals, such programs can serve as a model for how a coherent substitution strategy can lead to tangible security improvements.

Note that the committee does not seek to prescribe any particular policy in this regard. There could be a variety of mechanisms for implementing the strategy, ranging from voluntary to mandatory, and from top-down to bottom-up approaches. The committee leaves the precise contours of a policy for the appropriate parties to develop, preferably in a collaborative fashion, but merely recommends that there should exist a strategy related to chemical substitution, with defined goals and stakeholders.

> FINDING 5-3: Despite ongoing industry practice and some initiatives that previously operated under DHS's CFATS program, the strategy documents that were made available to the Committee do not cite chemical substitution as a key part of an overall chemical security strategy.

> **RECOMMENDATION 5-3: Substitution of safer alternative chemicals for hazardous chemicals in industrial and academic settings should be included as part of the overall strategy. The planning and development of these strategies should be spearheaded by DHS's Chemical Information Sharing and Analysis Center under a reauthorized CFATS[17] and should continue to be conducted in conjunction with regulatory agencies, specifically, the EPA, OSHA, as well as representatives from industry and academic research environments.**

---

[17] At the time of writing this report, the statutory authority for the CFATS program (6 CFR Part 27) expired and has yet to be reauthorized.

5.1.4c Insider Threat

Given the nature of insiders' privileged access to and knowledge of a facility's systems, an insider with malicious intent poses a special threat to any facility. The subject of "insider threat" has garnered significant attention and mitigation efforts in the past 15 years (Bunn and Sagan, n.d.). There have been substantial efforts to develop general insider threat detection and mitigation programs (FEMA, n.d.), particularly within government agencies and for defense contractors (DHRA, n.d.). Yet, in certain sectors—often related to the materials consumed or produced therein—the threat lies not only in the theft of information and the disruption of an organization's functions, but also in the possibility that sabotage by insiders could have extremely detrimental consequences for broader public health and safety. The release of over 40 tons of highly toxic methyl isocyanate from the Union Carbide insecticide plant in Bhopal, India, in 1984 and the thousands of resulting deaths provide an indication of the scale of harm that could result from a major accident at a chemical facility, whatever the cause (Broughton, 2005). Therefore, sabotage of facilities and transportation vehicles containing highly toxic chemicals poses a serious threat, particularly as it eliminates the need for a terrorist to procure the chemical of concern in quantities sufficient to create massive death and destruction by a deliberate release, as was suspected in the Bhopal disaster (Broughton, 2005).

The worst incident of accidental chemical release to date was the 1984 Bhopal disaster. On the night of December 3, 1984, 40 tons of methyl-isocyanate (MIC) gas were accidentally released from a pesticide plant in Bhopal, India. Environmental conditions were particularly favorable for dispersion, and a plume of gas dispersed over the city of approximately one million inhabitants. Although there remain many uncertainties regarding the consequences of the release, it is likely that as many as 500,000 people were exposed to the gas. At least 3,800 people died immediately, most from an impoverished neighborhood in immediate proximity to the plant, with several thousand more dying over the next few days. An unknown, but almost certainly very large, number of people suffered minor to severe injuries, including permanent blindness, some portion of which resulted in premature deaths. Although there have been allegations that the release was the result of a deliberate act by a disgruntled employee, these claims have lacked credence and ultimately the disaster appears to have resulted from mismanagement combined with inadequate training and mechanical failure (Broughton, 2005).

Yet, despite the sizable number of facilities housing dangerous chemicals, there is no explicit reference in any of the strategy documents surveyed to addressing the insider threat in the chemical terrorism context. This is in contrast to the direct mention of the challenge posed by insiders in the context of nuclear and biological weapons within United States strategy documents.[18] It is also worth noting that in the case of Bhopal and at least one other case, company representatives and others attempted to place blame on saboteurs or terrorists when courts concluded that willful negligence was the cause of an explosion.[19]

---

[18] See p. 9 of the National CWMDT Strategy.

[19] Union Carbide claimed that the Bhopal disaster resulted from sabotage by an employee, but the courts

Although safety and security are the same word in many languages, safety is not synonymous with security. OPCW defines safety as "measures to prevent *non-deliberate* releases of toxic chemicals into the environment and to mitigate the impact if such events occur" and security as "measures to prevent *deliberate* releases of toxic chemicals and to mitigate the impact if such events occur" (OPCW, 2016. Pg.3).

Insider threats therefore focus on concerted efforts of someone with access and knowledge to circumvent safety systems and deliberately release hazardous chemicals with the intent of creating harm. It should be noted that neither the Responsible Care Security Code of the American Chemistry Council nor the OPCW convention single out insider threats. The committee did learn that the CFATS program included some practical efforts to counter insider threats within the chemical industry (Gotten, 2020). For example, CFATS required criminal background checks of personnel working at high-threat facilities. Their staff could facilitate the vetting of personnel in regulated facilities against Terrorist Screening Databases, and CFATS provided some training materials that touch on insider threats. However, mitigating insider threat goes far beyond such basic checks and the assistance program appeared to be quite limited and nascent. The National Insider Threat Task Force under the ODNI does not have any programs that specifically apply to the chemical sector, nor was the committee able to find evidence of similar efforts to the former CFATS activities directed towards government facilities or within academic research institutions.

The complexity of the chemical industry and chemical facilities in other sectors is likely to increase as the industry grows and the integration of cyberphysical systems in chemical manufacturing and transportation introduce additional vulnerabilities. The unique character of this industry requires dedicated research to understand the insider threat in this context and then a focused program to develop tools and procedures to mitigate the threat. This distinct effort, which would need to be resourced, would not be starting from scratch, but can fruitfully build on existing insider threat expertise, and potentially draw on best practices from both the nuclear and biological security domains. Such a program would be most useful if it were a collaborative private-public effort, most likely between CISA, the National Insider Threat Task Force and the chemical industry. One potentially fruitful approach for implementation would be for CISA to expand the assistance it can provide industry both through former CFATS channels (see Box 5-3 for description of CFATS' previous activities) and to other facilities through the emerging ChemLock[20] program.

At the same time, there are considerable ongoing government and nongovernment efforts in the nuclear, radiological, and biological domains particularly designed to mitigate the insider threat in these specific contexts.

---

concluded that the disaster was caused by negligence. (Eckerman, 2005; Indian Express, 2017). Seventeen years later, an explosion in Toulouse was blamed on terrorists (France 24, 2017).

[20] CISA's ChemLock program provides risk evaluations and solutions for facilities that possess dangerous chemicals (CISA, 2024).

FINDING 5-4a: The strategic documents surveyed do not explicitly mention insider threat in the chemical terrorism context.

FINDING 5-4b: While CFATS included some practical efforts to counter insider threats within the chemical industry, the scope of these efforts appears to be limited and they have been discontinued, and the committee did not find evidence of a similar systematic program, either directed towards government facilities or within academic research institutions.

CONCLUSION 5-4: The significant potential consequences of an insider at a chemical facility conducting or assisting an attack warrants explicit inclusion in existing strategies and comprehensive policies to counter insider threats at any facility containing significant quantities of toxic chemicals.

RECOMMENDATION 5-4: Counter-insider-threat activities should be incorporated explicitly into broader CWMD strategy. ODNI, with other public and private partners, should develop a strategy to ameliorate insider threats explicitly for the chemical domain.

5.1.5 Examples of Current Program Activities Not Mentioned in Strategies

The committee found some cases in which important activities that the USG is appropriately undertaking are surprisingly absent from the strategy documents reviewed. Here are four significant examples:

On the interdiction side, the strategy documents do not explicitly mention developing military capabilities to recognize early warning signs of chemical terrorism plots, despite the committee being aware that such activity occurs in practice.

In contrast with the military, there is no explicit mention in the strategies of domestic LE possessing the ability to counter chemical threats at the tactical level, including containment, disablement, and neutralization. While hazardous materials (HAZMAT) first responders and some specialized divisions of the FBI possess some of these capabilities, they are not referenced in the strategies. This observation is in contrast with such references with regard to nuclear/radiation, for example.

The committee found only a few and indirect references to integrating combating WMD terrorism, and specifically chemical terrorism, within broader counterterrorism efforts. Combating WMD, including chemical terrorism, requires some specialized activities and capabilities. At the same time, chemical terrorism is a subset of the broader terrorist threat, and preventing and countering chemical terrorism must therefore integrate closely with other, non-WMD, aspects of the chemical terrorism threat. Beyond a general reference to incorporating WMD-specific considerations into intelligence activities (which amounts in the current instance to technical chemical capabilities and knowledge) and collaboration between chemists/chemical weapons experts and the IC, there appears to be insufficient mention in the strategies of employing or integrating with broader chemical terrorism prevention and countering approaches.

The committee is aware of some integration in practice, and the particulars are

BOX 5-3
Congressional Actions to Address Chemical
Threat through Sabotage of Facilities

Mindful of the threat posed by chemical facilities, Congress included in the Department of Homeland Security Appropriations Act of 2007, Pub. L. No. 109-295, a section (§550) which directed the Secretary of the Department of Homeland Security to promulgate "interim final regulations establishing risk-based performance standards for security of chemical facilities" within six months of the enactment of the Act. This Act also mandated the development of vulnerability assessments, as well as the development and implementation of site security plans for high-risk chemical facilities. "Facilities" include critical infrastructures that use, manufacture, store, or handle specific quantities of chemicals that DHS has identified as being extremely dangerous. The Chemical Facility Anti-Terrorism Standards (CFATS) regulatory program was developed to fulfill the requirements of the 2007 Act, but authorization for the program was not renewed in 2023. Previously managed by the Cybersecurity and Infrastructure Security Agency (CISA), this program identified and regulated high-risk chemical facilities to ensure security measures were in place to reduce the risk of certain hazardous chemicals being weaponized.

In 2014, Congress enacted the Protecting and Securing Chemical Facilities from Terrorist Attacks Act of 2014, Pub. L. No. 113-254, to reauthorize and codify the CFATS Regulatory Program (6 U. S. C. ~<([A-Z]).([A-Z]).|U. S. §§ 621-29). This legislation laid the foundation for the continued maturation of the CFATS program, adding new provisions as needed while preserving most of the existing regulations. Subsequent legislation in 2019 and 2020 extended the program until July 2023. The overall regulatory program for chemical facility and transportation defense, as administered through CFATS and other policies involved:

- **Identification:** Identifying high-risk facilities in key infrastructure sectors: CFATS used a dynamic multitiered risk assessment process to identify facilities that were high risk.
- **Security Plans:** Required the facilities judged to be at high-risk tiers to develop and implement appropriate Security Plans that met applicable risk-based performance standards (RBPS) for securing chemicals of interest. These standards included ensuring that they had effective security measures in place, so the risks associated with these chemicals could be mitigated. Security plans included items such as employee screening and background checks, area perimeter fencing, intruder detection systems, restricted access, video-verified monitoring, rigid maintenance systems, and employee training.

continued

BOX 5-3 Continued

- **Verification:** Inspections of facilities to validate the implementation of approved Security Plans were carried out by CISA,
  which was authorized to conduct inspections and enforce the
  provisions. CISA inspectors had the authority to enter, inspect,
  and audit the property, equipment, operations, and records of
  CFATS-covered facilities.

---

difficult to assess in the unclassified setting. The same can be said for integrating efforts to prevent/counter chemical terrorism into broader efforts to counter smuggling (which involves different sets of government actors).

When the strategies refer to multilateral activities, they solely reference the CWC and its associated implementing agency, the OPCW. These are the highest-level focal points for international efforts to address chemical weapons threats, but they are primarily focused on state-level threats as opposed to terrorist ones and they are far from the sole contexts in which these threats are addressed. For example, other multilateral (involving most states) or plurilateral (involving many states) efforts include UN Security Council Resolution 1540 Committee, the Global Partnership Against the Spread of Weapons and Materials of Mass Destruction, the Proliferation Security Initiative, the Australia Group, and others (NTI, 2013; United Nations Office on Drugs and Crime, n.d.; United Nations Counter-Terrorism Committee Executive Directorate, n.d.).

> FINDING 5-5: The committee found that certain key activities that the U.S. government is appropriately undertaking were surprisingly absent from the strategy documents reviewed, including military capabilities to provide early warning of chemical terrorism plots; law enforcement capabilities to counter chemical threats tactically; integration with broader counterterrorism and countersmuggling efforts; and involvement with other multilateral activities beyond the Organization for the Prohibition of Chemical Weapons.

CONCLUSION 5-5: The committee concludes certain key activities that are undertaken in practice to prevent and counter chemical terrorism are sufficiently important to merit inclusion in strategy documents. The absence of such activities from the strategies could impact policy implementation, including budgeting, program prioritization, and other consequences. Including these activities in existing strategies would bolster the comprehensiveness, and therefore effectiveness, of existing strategies.

RECOMMENDATION 5-5: Agencies should work to reconcile operational practice with policy by supplementing extant strategies to include current omitted effective activities and programs for countering chemical terrorism. This

would ensure that effective practices are maintained, properly resourced, and reflected in comprehensive strategies.

5.2 IMPLICATION OF NATIONAL STRATEGIC SHIFT FROM VEOS TO GPC FROM THE PERSPECTIVE OF "PREVENT/COUNTER"

The shift in emphasis from threats posed by VEOs to state-sponsored threats arising from GPC (Caves and Carus, 2021)[21] further complicates the prevent and counter task. Notably, it is likely that resources for identifying VEO terrorism threats will be redirected toward GPC threats. Yet, if great power conflict intensifies, that has many potential implications for chemical terrorist threats, beyond the possible reduction of resources available to address them.

First, even if states simply maintain defensive programs against state-based chemical threats, as many do today, those programs have dual-use implications for (nonstate) offensive threats. That means that expertise, technology, and materials, including chemical agents, might illicitly migrate from defensive programs to nonstate actors, as apparently occurred in the biological weapons domain with the so-called Amerithrax case (DOJ, 2010). Second, states might also choose to engage in offensive chemical weapons activities—as some states, notably Russia, appear to be doing today—and technology, materials, expertise, and/or chemical agents might be illicitly transferred from those programs to nonstate actors either intentionally or otherwise. Third, states, including great powers, even if there is no leakage from their own programs, might use or support nonstate actors in acquiring or deploying chemical weapons. Witting or unwitting involvement of states in nonstate chemical terrorism could dramatically increase the sophistication of such attacks, including the agent employed and/or the means by which it is delivered. Finally, states, including great powers, might engage in the use of chemical agents in ways that might be categorized as "state terrorism"—as some have alleged both Russia and North Korea to have done with targeted attacks in recent years—though the committee recognizes that this concept is controversial and difficult to clearly define.

5.3 SUMMARY

The committee's assessment of unclassified strategies reveals opportunities to enhance direct deterrence of chemical terrorism by incorporating it into existing CWMDT terrorism strategies. A strategic communications effort that leverages preventive and mitigating measures against chemical terrorism as part of a policy of deterrence by denial was lacking in the evaluated documents. Chemical substitution, also, was not included in the overall chemical security strategy, despite ongoing industry practices and some initiatives under DHS's CFATS. Insider threats stemming from chemical facilities are a major prevent/counter concern; however, it was not explicitly addressed in the

---

[21] 2018 National Defense Strategy of the United States of America: Sharpening the American Military's Competitive Edge, December, 2017.

surveyed reports. This absence necessitates explicit inclusion in existing strategies and comprehensive policies. Finally, key activities—such as military capabilities for early warning, LE to counter chemical threats, and involvement in multilateral activities— were minimally discussed in the reviewed documents, even though they do occur in the USG based on briefings presented to the committee. The discrepancy between de jure vs. de facto strategies could affect chemical terrorism policy comprehensiveness. Including these activities would further ensure that effective practices are properly maintained and resourced. The next chapter examines strategies related to responding to chemical incidents whether 1) from an attack that was not prevented/countered or 2) from an accident.

REFERENCES

Becker, P., and M. Nalabandian. 2022. DTRA Cooperative Threat Reduction (CTR) Program, briefing to the Chem Threats Committee. August 8, 2022. ISN briefing.

Broughton, E. 2005. "The Bhopal Disaster and its Aftermath: A Review." *Environmental Health* 4(1): 6.

Bunn and Sagan, eds. n.d. National Insider Threat Policy. https://www.dni.gov/files/NCSC/documents/nittf/EO_13587.pdf; https://www.dni.gov/files/NCSC/documents/nittf/National_Insider_Threat_Policy.pdf.

Caves, J and S. Carus. 2021. The Future of Weapons of Mass Destruction: An Update. NIU Press.

CISA, 2024. ChemLock. 2024. https://www.cisa.gov/resources-tools/programs/chemlock.

Croddy, E. 2002. "Dealing with Al Shifa: Intelligence and Counterproliferation." *International Journal of Intelligence and CounterIntelligence* 15(1): 52–60.

DHRA (Defense Human Resources Activity). n.d. Insider Threat Program for Industry. https://www.cdse.edu/Training/Insider-Threat/; https://www.dhra.mil/perserec/; https://www.cdse.edu/Portals/124/Documents/jobaids/insider/insider-threat-job-aid-for-industry.pdf.

DOJ (The United States Department of Justice). 2010. Amerithrax Investigative Summary. https://www.justice.gov/archive/amerithrax/docs/amx-investigative-summary.pdf.

DOS (U.S. Department of State). 2003. Proliferation Security Initiative. https://www.state.gov/proliferation-security-initiative.

DOS, 2022. UNSCR 1540. Fact Sheet. 2022. https://www.state.gov/remarks-and-releases-bureau-of-international-security-and-nonproliferation/unscr-1540-2/.

Eckerman, I. 2005. *The Bhopal Saga—Causes and Consequences of the World's Largest Industrial Disaster*. India: Universities Press.

Executive Office of the President. 2002. National Strategy to Combat Weapons of Mass Destruction.

Exec. Order No. 13650. 2013. Executive Order on Improving Chemical Facility Safety and Security. 2013.

Exec. Order No. 13650. 2018. Final Amendments to the Risk Management Program (RMP) Rule.

FEMA (Federal Emergency Management Agency). n.d. Protecting Critical Infrastructure Against Insider Threats. https://www.dni.gov/index.php/ncsc-how-we-work/ncsc-nittf/ncsc-nittf-resource-library; FEMA —Emergency Management Institute (EMI) Course I IS-915.

France 24. 2017. Seventeen Years Later, an Explosion in Toulouse Was Blamed on Terrorists. https://www.france24.com/en/20171031-france-total-subsidiary-found-liable-deadly-2001-azf-plant-blast-toulouse.

Gotten, F. 2020. Chemical Facility Anti-Terrorism Standards. Congressional Research Services. https://sgp.fas.org/crs/terror/IF10853.pdf.

Indian Express. 2017. "In Court, Defense Names Former Employee as 'Saboteur.'" https://www.bhopal.com/document/case-study.pdf#xd_co_f=ZmEwNjFkY2EtZDM5My00OGRjLThhOTMtNTdmODE0MDk1NDky.

Kaszeta, D. 2018. Why Are Nerve Agents so Difficult to Make? Bellingcat. https://www.bellingcat.com/resources/articles/2018/08/13/nerve-agents-difficult-make.

Lieberman, J., and M. Itamura. n.d. Cesium Irradiator Replacement Project—A Case Study. Sandia National Laboratory. https://www.nationalacademies.org/documents/embed/link/LF2255DA3DD1C41C0A42D3BEF0989ACAECE3053A6A9B/file/DF5A515DBB130AA64EC2C3EC7BAB52A4DDAEED6919CD?noSaveAs=1.

McNelis, P. 2022. FBI Counter WMD Intelligence Analysis Section, Briefing to the Committee.

NTI (Nuclear Threat Initiative). 2013. Proliferation Security Initiative (PSI). http://www.nti.org/treaties-and-regimes/proliferation-security-initiative-psi/.

OPCW (Organisation for the Prohibition of Chemical Weapons). n.d.(a). https://www.opcw.org/our-work/promoting-chemistry-peace.

OPCW. n.d.(b). https://www.opcw.org/our-work/preventing-re-emergence-chemical-weapons.

OPCW. 2016. Report on Needs and Best Practices on Chemical Safety and Security Management. https://www.opcw.org/sites/default/files/documents/ICA/ICB/OPCW_Report_on_Needs_and_Best_Practices_on_Chemical_Safety_and_Security_ManagementV3-2_1.2.pdf#13.

OSHA (Occupational Safety and Health Administration). n.d. Transitioning to Safer Chemicals — Why Transition to Safer Alternatives? https://www.osha.gov/safer-chemicals/why-transition.

Savage, T. 2022. FBI Policy Program, Briefing to the Committee.

Sharp, S. 2022. FBI ChemBio Countermeasures Unit, Briefing to the Committee.

Sherwood-Randall, E. 2023. The Biden Administration's New Strategy for Countering WMD Terrorism. Nuclear Threat Initiative.

Subcommittee on Cybersecurity and Infrastructure Protection. 2018. Committee of Homeland Security, House of Representatives, 115th Congress, Second Session. Industry View of the Chemical Facility Antiterrorism Standards Program, Hearing. February 15, 2018, Serial # 115–149, p. 32.

The White House. 2023. FACT SHEET: President Biden Signs National Security Memorandum to Counter Weapons of Mass Destruction Terrorism and Advance Nuclear and Radioactive Material Security. https://www.whitehouse.gov/briefing-room/statements-releases/2023/03/02/fact-sheet-president-biden-signs-national-security-memorandum-to-counter-weapons-of-mass-destruction-terrorism-and-advance-nuclear-and-radioactive-material-security.

UNODC (United Nations Office on Drugs and Crime). n.d. Countering Chemical, Biological, Radiological and Nuclear Terrorism. UNODC. https://www.unodc.org/unodc/en/terrorism/expertise/countering-chemical-biological-radiological-and-nuclear-terrorism.html.

UN Office of Counter-Terrorism Committee Executive Directorate. n.d. Ensuring Effective Interagency Interoperability and Coordinated Communication in Case of Chemical and/or Biological Attacks. 2017. https://www.un.org.counterterrorism/files/uncct_ctitf_wmd_wg_project_publication_final.pdf.

UNSC (United Nations Security Council). 2004. Security Council resolution 1540. Concerning Weapons of Massive Destruction. S/RES/1540. https://www.refworld.org/docid/411366744.html.

6

Adequacy of Strategies to Respond to Chemical Terrorism

Summary of Key Findings, Conclusions, and Recommendations

FINDING 6-1: The current compilation of U.S. strategies, operational plans, resources, and interagency agreements has yielded a network of first responder communities capable of robust response to most industrial and transportation chemical incidents regardless of their cause. Existing chemical accident first responder capabilities (e.g., for industry and transportation) are also useful for chemical terrorism scenarios.

FINDING 6-2: Tools (e.g., Wireless Information System for Emergency Responders (WISER, ACTKNOWLEDGE) that seek to bridge and enable better chemical, biological, radiation, and nuclear (CBRN) response communication across federal, state, local, territorial, and tribal (SLTT) organizations have been discontinued.

FINDING 6-3: The U.S. Global Deterrence Framework and other strategies involving the whole of government sharing often include representatives from the first responder and export control communities. This inclusion ensures that the USG will receive timely, up-to-date threat assessments and can make changes to their tactics, techniques, and protocols.

CONCLUSION 6-4: The National Response Framework (NRF) has adequately addressed chemical terrorism categorically under the Emergency Support Function #10: Oil and Hazardous Materials Response.

Summary Continued

FINDING 6-5: With respect to responding to chemical terrorism, the hierarchy of U.S. strategies, frameworks, and other guidance is complex; accordingly, their translation into operational practice may be challenging.

FINDING 6-6: With respect to the NRF, the first response communities, civil defense organizations, Department of Homeland Security (DHS), Department of Defense (DoD), and medical communities are continuing to exercise communication channels and are bringing awareness of such channels to relevant users. The number of potential venue targets is vast and response exercises simulating chemical attacks are being integrated into doctrine to provide experience and information to as many SLTT responders as possible.

RECOMMENDATION 6-6: Considering the complexity of the chemical threat space and U.S. government (USG) coordination required for an effective response to a chemical event, the committee recommends continuing a robust program of interagency exercises and trainings that practice communication and resource sharing.

The United States has well-defined authority and organizational constructs for emergency response, including large-scale and chemical terrorism response. The extensive multi-agency response capabilities of the United States are complexly governed, coordinated in policy, sufficiently connected to intelligence activities, and sufficiently capitalized; however, a mass casualty, multipoint, or cross-jurisdiction incident could have an impact beyond the SLTT capabilities. The committee has identified opportunities for improvements, but in the context of a great power competition (GPC)-focused national strategy, the committee found it difficult to recommend dramatic investments or changes. Maintenance, exercise, and integration of modernized response capabilities remain essential.

6.1 ANALYSIS OF STRATEGIES FOR "RESPONDING" TO WMDT CHEMICAL THREATS

Using our robust methodology (described in Chapter 3) the committee reviewed the strategy documents listed in Box 6-1 focusing on response to chemical terrorism. These documents contained highly variable content relating to the response aspects of combating chemical terrorism. The DHS and DoD documents were most useful for response. Most of the strategy documents espoused a coherent strategy or set of strategy elements (i.e., comprising a combination of a well-defined goal with a corresponding definition of success, as well as at least one policy, plan, and/or resource allocation designed to meet the goal(s)). The exception to this was DoD Directive 2060.02, which

---

BOX 6-1
Strategy Documents Reviewed by
Committee for "Response" Analysis

1. The National Strategy for Countering WMD Terrorism
2. The DHS Chemical Defense Strategy
3. Chemical and Biological Defense Program (CBDP) Enterprise Strategy
4. DoD Strategy for Countering Weapons of Mass Destruction (2014)
5. JP 3-40: Joint Countering Weapons of Mass Destruction

---

did not provide clear definitions of success for its goals of "Dissuade, deter, and defeat actors concern and their network[. . .]. Manage [weapons of mass destruction] WMD terrorism risks from hostile, fragile or failed states and safe havens.[. . . Or] [l]imit the availability of WMD-related capabilities."

With respect to whether the existing strategies, as encompassed by the above-mentioned strategic documents, are adequate to address both the current and emerging threat of chemical terrorism, most of the elements that we believe are essential to accomplish this task are reflected in the strategies. The matrix described in Chapter 3 (page 62) indicates these elements and whether or not each is addressed by the strategies.

6.1.1 Committee Definition of Adequacy: Response

For this study, response is defined as, "in countering weapons of mass destruction, the activities to attribute responsibility for an event: minimize effects, sustain operations, and support follow-on actions."

In the opinion of the committee, the concept of adequacy for strategies for responding to chemical terrorism will include elements that sufficiently address the following questions:

1. Does the U.S. strategy adequately enable response capabilities (e.g., operations coordination, information-sharing, medical support, and others) that minimize potential impact on life, property, and the environment?
2. Is the strategy for responding to chemical terrorism and are the resources devoted to implementing the strategy aligned with the priorities of the United States (e.g., protecting the homeland, ensuring economic security, maintaining military strength) and aligned with the nation's risk posture?
3. Does the strategy anticipate emerging threats by suggesting the scientific research and interagency relationships necessary to respond to future threats?

Overall, the committee found the strategies to be at least adequate, specifically, *DHS Chemical Defense Strategy, CBDP Enterprise Strategy, and JP 3-40: Joint Countering Weapons of Mass Destruction.*

6.1.2 Response Capabilities: Known vs. Unknown Threats

The committee believes that key elements to a prompt, effective first response are rapid availability of **situational awareness, technical information,** and **physical assets.** Hence, the focus is on the adequacy and timeliness of the communication chain as well as the adequacy of the content of the information conveyed by federal agencies to the first responders during a chemical event. After reviewing multiple briefings and evaluating the U.S. strategy documents, the committee observed that several federal agencies could be involved with information flow to first responders before, during, and after a chemical event, either accidental or terrorist. These agencies include DHS/Federal Emergency Management Administration (FEMA), DHS/Cybersecurity and Infrastructure Security Agency (CISA), International Association of Fire Chiefs (IAFC), Federal Bureau of Investigation (FBI), Drug Enforcement Administration (DEA), Environmental Protection Agency's (EPA's) NRP, local authorities, and others.

The challenges and adequacy of response to a chemical weapon of mass destruction (CWMD) incident vary greatly depending on whether the incident involves known or unknown threats.

Known, Existing Threats

It should be noted that the adequacy of a successful response that minimizes the effects of such an event is a function of the adequacy of Emergency Preparedness. To that extent, response to a chemical event at an existing threat location is more manageable as the nature of the threats (chemicals) are known, risks have been clearly identified and mitigated as much as possible, and the response teams are known (appropriate SLTT responders). Often the response teams also have experience in addressing these threats through regular training, tabletop drills, and exercises. Further, since the requisite response resources are well known to the responders, staging of response, countering resources, and equipment can be preplanned and made readily available during the event.

Unknown Threats

However, a chemical event anywhere else in the homeland, and without prior notice, presents challenges as the chemical nature and the amounts are unknown and therefore the risks to life are unknown at the outset. The immediate success of response in this case depends primarily on the first responders at the SLTT levels and their ability to rapidly reach out for appropriate additional resources when necessary. It should be noted that the U.S. Military response is not automatically triggered by a chemical incident (see Appendix A, USG Strategy Documents Provided to the Committee). Further, the USG response would only be required when SLTT authorities are overwhelmed or require specialized expertise (see page 111, USG Strategy Documents).

The *2018 National Strategy for Countering Weapons of Mass Destruction Terrorism* (White House, 2018) updated several approaches including strengthening outreach to responders by establishing lines of communication with federal agencies that greatly improve coordination before, during, and after an event. It also stated that providing training and equipment to SLTT entities will be continued with the aim of "creating self-sustaining capabilities that are not continually dependent on Federal assistance." (White House 2018). This places a burden on the first responders who have varying degrees of skills and resources and elevates the importance of training, education, availability of response resources, and a well-rehearsed and thoroughly familiar chain of communication and command.

6.1.3 Accidental or Intentional Chemical Incidents

In addition to understanding types of threats (unknown or known), whether an event is accidental or intentional factors into an adequate response. Approximately 800,000 hazardous shipments move every day in the United States, which equates to more than 3 billion tons of hazardous materials transported every year. During these material movements, more than 25,000 hazardous materials (hazmat) incidents occurred, which in the period of 2012–2022, caused less than 100 recorded fatalities and less than 2000 injuries (DOT, 2023). When accidents occur, first responders have tools, training, and interagency agreements generally adequate for protecting the U.S. population, themselves, and the environment. Specifically, the EPA's Emergency Support Function (ESF) #10-Oil and Hazardous Materials Response states that the EPA will provide: *federal support in response to an actual or potential discharge and/or uncontrolled release of oil or hazardous materials when [ESF is] activated. (p 1)*

The EPA is the primary agency that coordinates support from several other agencies including the Department of Agriculture, Department of Commerce (DOC), DoD, Department of Energy (DOE), Department of Health and Human Services (DHHS), DHS, Digital Object Identifier (DOI), Department of Justice (DOJ), State Department, Department of Transportation (DOT), General Services Administration, and Nuclear Regulatory Commission (EPA, 2008). The thirteen support agencies should contain the expertise necessary for the breadth of chemical incidents and the ability to reach out when additional resources or knowledge is needed.

The DHS 2019 Chemical Defense Strategy treats response to chemical terrorism and accidental release equivocally. The document states: *The Nation faces a complex threat landscape, especially from the evolving nature of the chemical threat, whether from accidental release or terrorist attack.*

A chemical terrorism event involving chemical weapons could pose challenges beyond the technical capabilities of first responders to promptly recognize or mitigate. The Joint Chiefs of Staff, Joint Publication 3-41 Published September 9, 2016, "provides joint doctrine for military, domestic, or international response to minimize the effects of a CBRN incident." (Pg. 3). The fundamentals of the military's role in response to

WMD are covered in Joint Chiefs of Staff Joint Publication 3-40. The specific definitions and counts of chemical terrorism events vary. Per guidance from the study sponsor to the committee and the Statement of Task (SOT), the chemical terrorism incidents for the committee's focus are those directed at U.S. assets, continental U.S. (CONUS) or outside the contiguous U.S. (OCONUS) excluding those that are state-sponsored. None of the databases mentioned below categorize incidents in ways that are exactly aligned with committee's tasking, making meaningful comparisons among data sets difficult. The University of Maryland Global Terrorism Database (START, 2022) describes about 30 acts of terrorism in the United States involving chemicals over the 50-year period 1970–2020. Using different inclusion criteria, the profiles of incidents involving chemical, biological, radiological, or nuclear and nonstate actors (POICN) Database describes 68 chemical terrorism cases (and 36 uses) from 1990–2020. Despite difficulty comparing incident counts, and timeframes, chemical terrorism is historically a miniscule portion of chemical release events that require first responders.

In sum, the vast majority of chemical incidents in the United States are not terrorism; thus, the vast majority of first responder actions associated with chemical releases are from accidents, transportation incidents, or the results of natural phenomena—not responses to terrorist events. Thus, local first responders would respond to a chemical terrorism event even if the origin or motivation of the chemical release were unclear: accidental, sabotage, or terrorist. When necessary, intelligence assets are engaged.

> FINDING 6-1: The current compilation of U.S. strategies, operational plans, resources, and interagency agreements has yielded a network of first responder communities capable of robust response to most industrial and transportation chemical incidents regardless of their cause. Existing chemical accident first responder capabilities (e.g., for industry and transportation) are also useful for chemical terrorism scenarios.

6.1.4 Advance Detection Capabilities

DHS has recognized that to assess the impact of a chemical event in real-time, the agency would need chemical modeling programs, which would also provide valuable information to first responders—a key aspect to improving situational awareness and understanding physical assets. CISA created Jack Rabbit (DHS, 2021), which is a multiagency initiative (DHS, EPA, DoD) aimed at providing data and information related to chemical threats through field studies and experiments (e.g. laboratory experiments, wind tunnel experiments, and urban dispersion modeling). Jack Rabbit III, in particular, focuses on modeling tools and detection technologies to better understand and monitor chemical threats such as a large-scale ammonia release via dispersion (plume size, dispersion rate, ammonia concentration). Furthermore, the modeling is expected to improve the following:

- planning for release incidents;
- emergency response;

- mitigation measures to reduce the impact on affected populations and infrastructure;
- guidance and data for emergency response procedures; and
- validation of protective action distances.

Figure 6-1 is an example of using a portable gas detector, miniRAE, to measure and map the concentration levels of ammonia at varying distances from the chemical's point of release in real time. Coupling this capability with a communications network for first responders, such as FEMA's ChemResponders (ChemResponder Steering Committee, 2020), will support emergency response decision-making. Other advanced technologies—high-definition video recording equipment, drones, hyperspectral imaging technologies, and others—are being explored for real-time applications as a way to provide adequate information to increase first responder's safety on-site (e.g., using the appropriate protective equipment).

The committee also recognizes that information from chemical release modeling, such as Jack Rabbit, and the subsequent information flow to first responders is agnostic to the motivation behind the chemical release: terror, sabotage, or accident. If the chemical(s) released have known physical properties, such modeling information is likely to be more accurate than with an unknown.

Unit # (downfield distance from the release)	Ammonia Con. at 3:30 AM (Maximum Con.)
1 (0.1 km)	0.8 (0.8) ppm
2 (0.1 km)	9.9 (9.9) ppm
3 (0.1 km)	5.2 (7.7) ppm
4 (0.1 km)	1.0 (1.1) ppm
5 (0.1 km)	0.5 (0.6) ppm
6 (0.3 km)	0.4 (0.5) ppm
7 (0.3 km)	1.7 (2.4) ppm
8 (0.3 km)	0.6 (0.6) ppm
9 (0.5 km)	0.2 (0.3) ppm
10 (0.3 km)	0.2 (0.3) ppm
11 (0.2 km)	0.5 (0.5) ppm
12 (0.2 km)	8.4 (13.1) ppm

FIGURE 6-1 (Top) Photograph of miniRAE gas detection monitors used in field testing. (Bottom) Modeling of an ammonia release in real-time as a function of distance from release (km) and concentration (ppm). Concentration data was collected using miniRAE.
SOURCE: Fox and McMasters, 2021.

6.1.4a Military Preparedness: Identification or
Recognition of a Chemical Weapons Event

For many chemical terrorism events, identification of the chemical(s) involved is straightforward as readily available documentation such as site chemical inventories or shipping manifests can be used to identify the potential chemicals involved. Figure 6-2 shows the different types of chemical agents developed for military use, their chemical properties, and the signs and symptoms of a person who is exposed. Should a terrorist bring a chemical to the location of an attack, characterization of the threat could take longer and would be based on physical observations, observed symptoms, and monitoring technologies. While the committee believes chemical weapon use to be unlikely for a terrorism event, there is a history of chemical agent attacks (discussed below), and the consequences could involve significant loss of life in the first responder community.

The initial recognition or identification of chemical weapons in a terrorist attack may be slower than desired as first responders may not be trained in recognizing symptoms of chemical agent exposure. Without prior indication, the mindset of the first responders will not be focused on a potential chemical weapon attack. For example, the symptoms of a chemical attack may be similar to those of other medical conditions or chemical exposures. Exposure to a nerve agent can produce a seizure response, followed by cardiac arrest, as can multiple medical conditions. First responders commonly encounter such symptoms among the general population, but only extremely rarely if ever are those symptoms caused by nerve agent exposure.

This situation is in marked contrast to battlefield scenarios that are considered by the U.S. Army, where the use of chemical weapons is possible and may have a reasonable likelihood of occurring. Accordingly, the Army is well equipped to identify

	Nerve Agents		Blister Agents (injure skin, eyes, and airways)		Blood Agents (cause blood changes and heart problems)		Choking Agents	
Examples	Sarin	VX	Mustard	Lewisite	Hydrogen Cyanide	Cyanogen Chloride	Chlorine	Phosgene
Odor	Odorless		Garlic or Mustard	Geraniums	Burnt almonds		Bleach	Mown hay
Persistency*	Non-persistent (min. to hrs.)	Persistent (>12 hrs.)	Persistent		Non-persistent		Non-persistent; vapors may hang in low areas	
Rate of Action	Rapid for vapors; liquid effects may be delayed		Delayed	Rapid	Rapid		Rapid at high concentrations; delayed at lower concentrations	
Signs and Symptoms	Headache, runny nose, salivation, pinpointing of pupils, difficulty in breathing, tight chest, seizures, convulsions, nausea, and vomiting		Red, burning skin, blisters, sore throat, dry cough, pulmonary edema, eye damage, nausea, vomiting, diarrhea. Symptoms may be delayed 2 to 24 hrs.		Cherry red skin/lips, rapid breathing, dizziness, nausea, vomiting, convulsions, dilated pupils, excessive salivation, gastrointestinal hemorrhage, pulmonary edema, respiratory arrest		Eye and airway irritation, dizziness, tightness in chest, pulmonary edema, painful cough, nausea, headache	
First Aid	Remove from area, treat symptomatically. Atropine and pralidoxime chloride (2-PAM chloride), diazepam for seizure control		Decontaminate with copious amount of water, remove clothing, support airway, treat symptomatically		Remove from area, assist ventilations, treat symptomatically, administer cyanide kit		Remove from area, remove contaminated clothing, assist ventilations, rest	
Decontamination	Remove from area, remove clothing, flush with soap and water, aerate							

*How long a chemical remains at toxic levels

FIGURE 6-2 Effects and treatment of selected chemical weapons developed for military use. SOURCE: DHS, 2006.

the presence of a chemical warfare agent in the environment. Near real-time detectors have matured and are common on military equipment. For example, the MINICAMS gas chromatograph system has been in use for more than a decade and is deployed on Army platforms. However, civilian first responders may be unlikely to have access to this type of measurement equipment, and even if it was accessible, would not likely have the budget to procure devices, or the personnel bandwidth to accommodate training to enable operation.

First responders will be unable to understand the extent of contamination, a situation that will be exacerbated by the inability to conduct any meaningful surface characterization. This situation affects first responders' decision-making regarding appropriate personal protective equipment. Semivolatile or low-volatile agents will largely partition to surfaces in the exposed environment, which means that inhalation risks from these agents are decreased and that the risk of toxic exposure by dermal contact is elevated.

The experience of the first responders in the Skripal poisoning incident illustrates this point. The risks that are illustrated include likely dermal contact with A234 or something like it, producing extremely toxic responses in the Skripal case. First responder Nick Bailey was exposed, reportedly when wearing personal protective equipment (PPE) (see Appendix E for "Skripal Poisoning" case study).

The Army at the Chemical Biological Center (CBC) is currently engaged in developing spray reagents that can identify agent contamination on skin surfaces, which would likely work on other surfaces as well. However, the spray reagents may not be applicable to large areas, and furthermore are relatively early on in the research and development pipeline; i.e., it may be years before these technologies are available for characterizing contamination on skin in military environments, and even longer for civilian use.

6.1.5 First Responder Input

A major component for creating a robust strategy is to ensure critical information is collected and included from the first responder community. The committee recognizes that some initiatives, like Jack Rabbit, have included first responder input in their study of safety risks of transporting ammonia—an essential chemical as a fertilizer and in fertilizer production. The transportation sector faces a large safety risk in its role of shipping ammonia across the country: chemical hazard spills or attacks at handoff points. In these incidents, first responders are at the highest risk. For example, police may lack training and enter a scene of an accidental release, or responders may not have the correct PPE. To ensure that the safety needs of this community are met, Jack Rabbit's questionnaire asked for input on where first responders see security gaps around anhydrous ammonia (Figure 6-3, top) as well as what equipment their department would need to effectively respond to an ammonia emergency (Figure 6-3, bottom).

CISA also recently released the SAFECOM Nationwide Survey (SNS), as a way to collect data from target participants, to identify gaps, and inform the program's strategic priorities for improving the nation's emergency communication systems (CISA.gov, 2018). SNS is most interested in gathering information from emergency communication

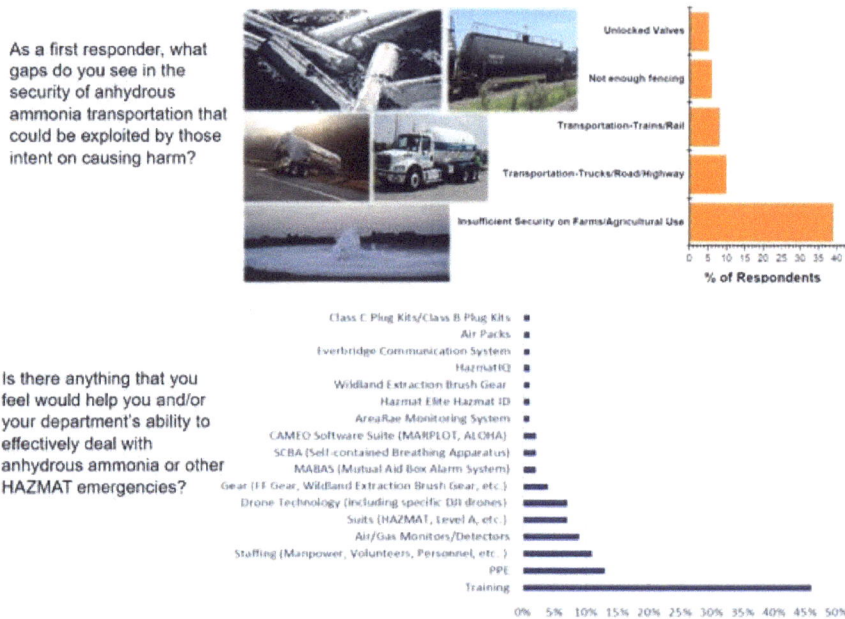

FIGURE 6-3 (Top) Survey results from first responders when asked, "As a first responder, what gaps do you see in the security of anhydrous ammonia transportation that could be exploited by those intent on causing harm?" (Bottom) Results from the same survey, where first responders answered the question: "Is there anything that you feel would help you and/or your department's ability to effectively deal with anhydrous ammonia or other HAZMAT emergencies?" The number of participants involved is not known.
SOURCE: Fox and McMasters, 2022.

centers, emergency management, law enforcement (LE), emergency medical personnel, fire and rescue professionals, and other organizations that use emergency communications technology to ensure public safety. In summary, these types of input from relevant stakeholders in the response community will also eventually shape the direction of risk assessments (e.g., type of chemical threat characterized, experimental setup, types of tools to develop) as will be described in Section 6.1.6.

6.1.5a Access to Intelligence

During briefings, the committee learned that it is possible that there is information available that would be most beneficial to the first responders and reduce causalities; however, it cannot be transmitted due to the classification status of the information (see Appendix A). It is recognized that the National Counterterrorism Center (NCTC) created a mobile app, ACTknowledge, that shares unclassified counterterrorism reports, analysis, training resources, and alerts to users. This app was created based on the recommendation of the 9/11 Commission as a way to integrate and share information related to strategic planning and government analysis across SLTT and federal partners.[1] However, as of January 2023, ACTknowledge was discontinued. The Office of the Director of National Intelligence (ODNI) also created a First Responder Toolbox, which is an ad hoc, unclassified, and For Official Use Only(FOUO) reference aid intended to promote counterterrorism coordination among federal, SLTT government authorities, and private sector officials to coordinate in deterring, preventing, disrupting, and responding to terrorist attacks. First Responder Toolbox could serve as a potential indicator of a chemical or biological attack (Joint Counterterrorism Assessment Team (n.d.).

Under the National Institutes of Health (NIH), the National Library of Medicine hosted the mobile app and web-based platform: Wireless Information System for Emergency Responders (WISER). This was designed to provide first responders with quick access to critical information during hazardous material incidents and other emergencies. The app included physical properties of the chemical, health effects, safety protocol, and other protective measures. It had emergency step-by-step response guides published by various agencies such as DOT. WISER was discontinued in February 2023, although NLM listed alternative publicly available sources on its website (NIH/ NLM, 2023).

The committee also recognizes that the FBI is actively engaged in fostering communication with state and local first responders, including the National Guard, and industry (Savage, 2022); however, it is not clear whether the outreach is comprehensive or systematic. The risk is that an event could occur in an area where first responders would not be aware of, or in communication with, FBI personnel or capability.

Overall, the committee assesses that there is not an obvious systematic means of communicating information to responders. Getting information to this community can be a significant challenge and represents a vulnerability.

FINDING 6-2: Tools (e.g., WISER, ACTknowledge) that seek to bridge and enable better chemical, biological, radiation, and nuclear response communication across federal and SLTT organizations have been discontinued.

FINDING 6-3: The U.S. Global Deterrence Framework, and other strategies involving whole-of-government sharing, often include representatives from the first responder and export control communities. This inclusion ensures that the USG will receive timely, up-to-date threat assessments and can make changes to their tactics, techniques, and protocols.

6.1.6 Risk Assessments

Risk assessments are important for steering the direction of strategies aimed at enhancing response to chemical terrorism. The EPA's Risk Management Program (RMP) was established to prevent and mitigate the consequences of chemical accidents. Critical facilities are required to periodically submit information to EPA that includes

---

[1] NOTE: ACTknowledge was discontinued in January 2023.

the facility's hazard assessment, accident prevention mechanisms, and emergency response measures. This plan provides local fire, police, and emergency response personnel with valuable information to prepare for and respond to chemical emergencies in their community.

Several programs within DHS develop different types of assessments at the local, state, and national levels. Every three to five years, FEMA releases the Hazard Identification and Risk Assessment (HIRA) and Threat and Hazard Identification and Risk Assessment (THIRA); both guidance documents address issues at the state and local levels, respectively.

Developments of THIRA, particularly in urban areas, effectively ask local communities to characterize risks and then use core capabilities to translate risks to required levels of capability. Planning committees, including firefighters, police, and paramedics, are asked to look across specific threats and decide on threat-agnostic capability and performance requirements. They are designed to drive a collaborative local planning process. Furthermore, when a jurisdiction assesses risks, they are explicitly considering the chemical terrorism risks, thus as THIRA evolves, appropriate attention to chemical terrorism would be given, especially in the context of chemical precursors.

Response communities are less engaged with the HIRA because these documents focus more on program development and evaluation at the state level. Additionally, the Probabilistic Analysis of National Threats and Risk Program (PANTHER) under CISA develops risk assessments often at the request of the FBI and other agencies. Within PANTHER, there is an emphasis on chemical risk assessment, much of which is at the FOUO level or higher.

At the national level, DHS and the intelligence community (IC) have developed the Strategic National Risk Assessment (SNRA). National-level risk assessments inform national strategies, just as THIRA would inform strategies at the state level. The linkages exist, but strategy ultimately comes from policy directions and prioritizations.

> Our risk assessments consistently show that even though nuclear and biological agent threats have the ability for catastrophic effects, the chemical threat has a much higher probability of occurring. So the chemical threat consistently shows up at or above the risk levels of the other threats but is consistently underfunded compared to the other threats. The chemical threat is not recognized for the risk that it poses (Fox, 2022).

6.1.7 Top-Down and Bottom-Up Information Flow

All of the briefings received by the committee from various agencies responsible for some part of antichemical terrorism activities (see list of briefers, Appendix A) demonstrated a clear understanding looking upstream to current authorities, strategies, policies, and laws governing internal agency responsibilities. Agencies understood their charter, authority, and responsibilities. **However, less clear to the committee is how the requirements systematically flow downstream from higher-level policy to subsidiary organizations and finally to first responders.**

For example, the committee identified that the roles and responsibilities of EPA and DHS/FEMA officials as well as their chains of communication are convoluted, which could lead to confusion at the local level. This lack of clarity could result in a slower response to a chemical incident or attack. The description in Box 6-2 highlights the complexity of understanding how strategies are deployed from a high-level (National Incident Management System [NIMS]) to the practitioner level: first responders.

The NRF prioritizes collaboration with the private sector and nongovernment organizations (NGOs), locally driven incident management, and active readiness to stabilize community lifelines and enable rapid and safe restoration of services in severe incidents.

The 4th edition of the NRF describes new initiatives that leverage existing networks and integrate business interests and infrastructure owners and operators into emergency management.

FEMA's approach has focused on training State and local responders by granting awards to enhance the capability of local response entities and issuing guidance to the local officials and first responders. FEMA notes that

> elected and appointed leaders in each jurisdiction are responsible for ensuring that necessary and appropriate actions are taken to protect people and property from any threat or hazard. When threatened by any hazard, citizens expect elected or appointed leaders to take immediate action to help them resolve the problem. Citizens expect the government to marshal its resources, channel the efforts of the whole community—including voluntary organizations and the private sector—and, if necessary, solicit assistance from outside the jurisdiction (FEMA, 2010, Pg. 13).

The NRF identifies various elected officers like governors, state emergency officers, and other agencies with various capabilities but does not define clear roles and responsibilities.

NRF also provides that in the event the state and local LE capabilities are overwhelmed by an attack incident, the DOJ is to assume the responsibility for coordinating federal LE activities to ensure public safety and security.

It is worth emphasizing that the NRF does not present any component of the Federal government—DHS, DoD, DOJ, or otherwise—as the prescribed owner or 'lead response agency' for any type of incident by default. ESF Annexes of the NRF lay out support functions that a federal agency may be called upon to assist (such as transportation, fire suppression, or energy). Similarly, the Incident Annexes of the NRF specify coordinating and cooperating federal agency roles within a narrow set of specific incidents. Neither Annex, however, supersedes the key principles of the NRF itself, which spell out a flexible, locally driven response concept, whose expansion to the federal "tier" occurs at the prompting of overwhelmed local and state officials.

CONCLUSION 6-4: The NRF has adequately addressed chemical terrorism categorically under the ESF #10: Oil and Hazardous Materials Response.

While biological and nuclear/radiological incidents have dedicated NRF Incident Annexes, incidents involving the release of a toxic chemical would ostensibly be captured by some combination of ESF#10 "Oil and hazardous materials response" (coordinated by the EPA), and the Terrorism Incident Law Enforcement and Investigation Annex (coordinated by the DOJ/FBI). **Because of this ambiguity, of all potential**

BOX 6-2
Relationship of Response and Preparedness Documents

The main framework employed by DHS/FEMA to coordinate and respond to emergencies, natural disasters, or terrorist events is the National Incident Management System (NIMS), which provides a common set of principles, practices, and procedures to facilitate incident management and response while maintaining the flexibility to address a breadth of incidents. It is designed to ensure interoperability and compatibility among different organizations involved in the emergency response; however, it is written at a high level. Within this framework, shown in Figure 6-1-1, are the National Planning Frameworks, where the NRF is located. Two documents related specifically to responding to chemical incidents are included under the NRF: ESF#10 and the Oil/Chemical Incident Annex.

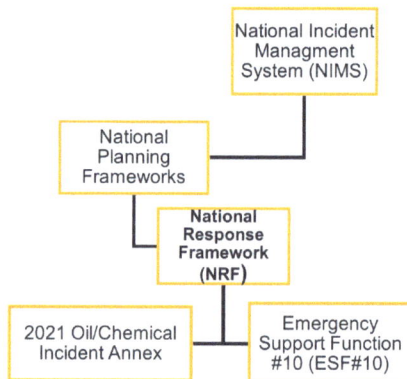

FIGURE 6-1-1 Diagram showing documents nested under the FEMA's NIMS. SOURCE: FEMA, 2021.

ESF #10, as described in Section 6.1.3, details federal support for an uncontrolled release of hazardous material. FEMA works in conjunction with other federal agencies, like EPA, and partners to coordinate ESF #10 activities during a chemical incident. Additionally, the ESF applies whether a presidential emergency has been declared under the Stafford Act (see Box 6-3), or not. Furthermore, the 2021 Oil/Chemical Incident Annex details important oversight, resourcing, laws, regulations, presidential directives, and federal response coordination. Although the word "chemical" is included in the title, sections of the Annex cover topics specific to oil spills, but it does not have sections specific to chemical spills or releases.

BOX 6-2 Continued

In summary, NIMS provides the overarching framework for incident management and coordination, while ESF #10 and the Annex operate within the NIMS structure to address specific hazardous materials incidents, including chemical incidents. The mechanisms employed are coordination, technical expertise, and support in collaboration with other response entities (i.e., first responders).

SOURCE: FEMA, 2021.

BOX 6-3
Stafford Act

The Stafford Act, officially known as the Robert T. Stafford Disaster Relief and Emergency Assistance Act, is a United States federal law that was enacted in 1988. This statute provides the legal framework for the response to, and recovery from, natural disasters, acts of terrorism, and other catastrophic events.

This law authorizes the President of the United States to issue a declaration of a major disaster or emergency, which then enables the federal government to coordinate and provide assistance to SLTT governments, as well as certain private nonprofit organizations and individuals affected by the disaster or emergency. The assistance provided under the Stafford Act can include financial aid, grants, loans, and other forms of support to help with response, recovery, and rebuilding efforts.

FEMA plays a central role in implementing the provisions of the Stafford Act. FEMA works in collaboration with various federal, and SLTT agencies to coordinate disaster response and recovery operations, provide technical assistance, and administer financial assistance programs.

SOURCE: Robert T. Stafford Disaster Relief and Emergency Assistance Act, PL 100-707 (November 23, 1988). see https://www.fema.gov/robert-t-stafford-disaster-relief-and-emergency-assistance-act-public-law-93-288-amended.

WMD terrorism scenarios, chemical incidents appear to have the greatest potential for interagency confusion, particularly in the early response stages wherein an accidental hazmat release may be indistinguishable from an act of chemical terrorism.

The Chemical Security Analysis Center (CSAC) generates a wealth of information on chemical threats, however vertical communication downward to the first responders currently occurs on an ad hoc basis. Relationships have been developed with entities like the Center for Domestic Preparedness (part of FEMA), the National Association of Fire Chiefs, the Ammonia Production Association, the Railroad Transportation Association, and other organizations. The purpose of these relationships is to transmit information that first responders would need, to enable them to recognize a chemical terrorist event. An example of successful communication stems from the results of the Jack Rabbit exercises, which are widely communicated. However, there does not appear to be a systematic approach for communication with the broad spectrum of first responders. Consequently, if there is an incident, it can be well after the fact that CSAC is involved. This situation is exacerbated by the fact that it is difficult to transmit timely information to responders if the information is classified or sensitive (see section 6.1.5 for more details).

Overall, chemical terrorism events (and any CBRN event) involve authority flowing among agencies and complex incident characterization. Impediments to this flow could slow response, delay event characterization, increase casualties, and confuse crime scene preservation.

FINDING 6-5: With respect to responding to chemical terrorism, the hierarchy of U.S. strategies, frameworks, and other guidance is complex; accordingly, their translation into operational practice may be challenging.

6.1.8 Emergency Response Coordination

The committee learned from briefings that while information does flow down to first responders from individual federal agencies; coordination among the different organizations can be improved to ensure first responders receive the needed information. Table 6-1 describes several federal agencies and their respective programs that would benefit from more coordination.

In 2006, Congress too acknowledged that there was a need for stronger coordination and national leadership to address gaps in emergency responders' abilities to communicate across jurisdictions and functions. During that time, Congress authorized the establishment of the Emergency Communication Preparedness Center (ECPC) (Section 671, Pub. Law No. 109-295), also known as the "Post-Katrina Emergency Management Reform Act." ECPC is a federal interagency within CISA. Its strategic priorities include increasing efficiencies at the federal level through joint investment and resource sharing and improvements in alignment of strategic and operational emergency communications planning across levels of government. In reality, there appears to be a patchwork of multiple agencies involved in providing training and resources in an uncoordinated manner to local first responders thereby enhancing the risk of critical gaps.

Further, there is the Nationwide Communications Baseline Assessment (NCBA) to evaluate the nation's ability to communicate during a variety of response operations; there seems to be a lack of a clear and timely transmission pathway for critical information that needs to be provided to first responder during a CWMD event. In an exploration of the reasons for this, it was clear that there are too many bureaucratic barriers that block the transmission of much-needed information in a timely manner.

ECPC considers various public communications technologies such as Next Generation 9-1-1, land mobile radio, long-term evolution, and others as a way to align strategic and operational emergency communications (interoperable and operable) across the levels of government. As stated in its 2019 Annual Strategic Assessment (ECPC, 2021. Pg. 2),

> The ECPC works to address gaps in emergency communications and enables emergency response providers and relevant government officials to continue to communicate in the event of natural disasters, public health emergencies, acts of terrorism, other man-made incidents, and planned events.

CISA also developed the National Emergency Communications Plan (NECP), which is a strategic plan to strengthen and enhance emergency communications capabilities in the United States. NECP aims to maintain and improve emergency communications capabilities for emergency responders and serves as the nation's roadmap for ensuring emergency communications interoperability at all levels of government. This plan establishes a shared vision for emergency communications and assists those who plan for, coordinate, invest in, and use operable and interoperable communications for response and recovery operations. This includes traditional emergency responder disciplines and other partners from the whole community that share information during incidents and planned events.

With respect to addressing chemical attacks specifically, the WMD Strategic Group Consequence Management Coordination Unit (WMDSG's CMCU) coordinates with FEMA through its office CBRN office. Recently, this office was replaced by the Office of Emerging Threats (OET) (FEMA, n.d.), where CBRN responsibilities are still retained. This type of collaboration provides strategic advice and recommended courses of action for ongoing LE and counterterrorism operations. The FBI has designated WMD coordinators in its 56 field offices with the idea that building strong working relationships in place makes for a smoother response to a chemical incident. They routinely host WMD workshops to train first responders in recognizing the use of WMD during the initial stages of an incident. During these exercises, trained agents and biowarfare scholars share lessons learned from past events. These activities provide an advanced, hands-on understanding of the hazards posed by WMD and increase first-response preparations to handle a WMD incident. The WMDSG's CMCU serves as a link between FBI-led crisis response and FEMA-led consequence management (CM) operations. Interagency coordination is exemplified through this initiative. In sum, the establishment of the WMDSG leads to improved federal interagency coordination for WMD-related terrorist threats and incidents.

After a chemical terrorism event, the FBI plays a significant role in response since they are the lead organization for WMD investigation. Addressing the language in this committee's SOT, "responding to chemical terrorism incidents to attribute their origin" the FBI is well positioned organizationally to attribute origin through its Criminal Investigation Division, Directorate of Intelligence, Weapons of Mass Destruction Directorate (WMDD), Counterterrorism Division, and Counterintelligence Division (FBI.gov; U.S. Government Accountability Office, 2023).

As shown above, the FBI appears to have a strong response system and DHS also has a robust training program for responding to chemical attacks. These agencies make clear efforts to have resources and programs available for on-the-ground responders. Response exercises that are integrated into the overall exercise programs are one way to ensure robust capability. Continuing chemical exercises will strengthen hazards preparedness routine. Lastly, in a resource constrained environment, response exercises will remain a necessity because they frequently test the response-coordinated groups' ability to pivot and operate in a dynamic environment.

> FINDING 6-6: With respect to the NRF, the first response communities, civil defense organizations, DHS, DoD, and medical communities are continuing to exercise communication channels and are bringing awareness of such channels to relevant users. The number of potential venue targets is vast and response exercises simulating chemical attacks are being integrated into doctrine to provide experience and information to as many SLTT responders as possible.

> **RECOMMENDATION 6-6: Considering the complexity of the chemical threat space and USG coordination required for an effective response to a chemical event, the committee recommends continuing a robust program of interagency exercises and training that practice communication and resource sharing.**

6.1.9 Medical Counter Measures (MCM)

Initiatives for collaboration across federal agencies exist to address issues around medical countermeasures (MCMs). Figure 6-4 illustrates how Biomedical Advanced Research and Development Authority (BARDA) prioritizes chemical agents for which MCMs are needed and made available. First, the relevant agent is identified based on intelligence, level of accessibility, and information related to the agent's previous use by nonstate actors. Second, a judgment will be made on how to respond to the chemical incident. Factors such as realistic time to treat and long-term effects on survivors are considered. Third, BARDA will assess its current level of preparedness given current USG holdings, approved and available medical treatments, and available routes of administration. The results of these assessments are used to identify and address gaps in current preparedness.

The assessment of the military/USAMRICD and the NIH MCM development is reviewed, and a critical assessment of their goals, priorities, implementation plans, and progress with findings and gaps is presented. In 2006, the Chemical Countermeasures Research Program (CCRP) under NIH-NIAID established the Medical Chemical Countermeasures Against Chemical Threats (CounterACT) Program. Although MCMs

TABLE 6-1 Federal Agencies and Programs Involved in Response

Federal Agency	Program	Description
Department of Homeland Security (DHS)	Federal Emergency Management Agency (FEMA) Cybersecurity and Infrastructure Security Agency (CISA) Office of Health Affairs (OHA) Science and Technology Directorate (S&T) Chemical Security Analysis Center (CSAC)	FEMA coordinates and supports emergency response efforts, including those related to chemical terrorism incidents. CISA provides expertise and support for protecting critical infrastructure, including chemical facilities. OHA works to ensure preparedness and response capabilities for public health emergencies, including chemical incidents. S&T conducts research and develops technologies to enhance the response and recovery from chemical terrorism incidents. CSAC provides chemical threat information, design and execution of laboratory and field tests, and a science-based threat and risk analysis capability to create best response to chemical hazards.
Department of Justice (DOJ), and Federal Bureau of Investigation (FBI)	Bureau of Alcohol, Tobacco, Firearms and Explosives (ATF) Weapons of Mass Destruction Strategic Group Consequence Management Coordination Unit (WMDSG CMCU) Directorate of Intelligence Critical Investigative Division	FBI investigates and responds to chemical terrorism threats and attacks, working in coordination with other agencies. ATF addresses the illegal use, acquisition, and trafficking of chemicals, including those used for terrorism. WMDSG CMCU develops response plans, shares intelligence, and coordinates efforts to address WMD, including chemical terrorism threats and incidents through external partnerships. Directorate of Intelligence gathers, analyzes, and disseminates intelligence related to chemical terrorism threats to members of the IC, LE, and private sector. Critical Investigative Division conducts investigations into domestic terrorism, including chemical terrorism incidents and related activities.
Department of Defense (DoD)	U.S. Army Medical Research Institute of Chemical Defense (USAMRICD)	USAMRICD develops medical countermeasures to chemical threats; and trains and educates medical personnel for the management of chemical causalities.
Environmental Protection Agency (EPA)	Criminal Investigation Division Emergency Response Program (ERP)	Criminal Investigation Division investigates violations of environmental laws related to chemicals, including those with potential terrorism connections. ERP provides technical expertise and resources to respond to chemical incidents, including those involving terrorism.

continued

TABLE 6-1 Continued

Federal Agency	Program	Description
Department of Health and Human Services (HHS); National Institute of Health (NIH)	U.S. Army Medical Research Institute of Chemical Defense (USAMRICD) Defense Threat Reduction Agency (DTRA) Biomedical Advanced R&D Authority (BARDA) National Institute of Allergy and Infectious Disease (NIAID)	USAMRICD conducts research on chemical defense and provides medical support in response to chemical incidents, including those involving terrorism. DTRA develops and deploys advanced detection and response capabilities to counter chemical threats, including terrorism. BARDA develops and procures MCM that address the public health and medical consequences of CBRNincidents. NIAID develops medical countermeasures against infectious agents that could be used in chemical attacks.
Office of the Director of National Intelligence (ODNI)	The National Counter Terrorism Center (NCTC)	NCTC produces threat analysis, maintains the authoritative database of known and suspected terrorists, shares intelligence, and conducts strategic operational planning.
Department of Agriculture (USDA)	Animal and Plant Health Inspection Service (APHIS)	APHIS works to prevent and respond to chemical threats, including agroterrorism incidents that may involve chemical agents.

BARDA Chemical MCM Threat Prioritization Scheme

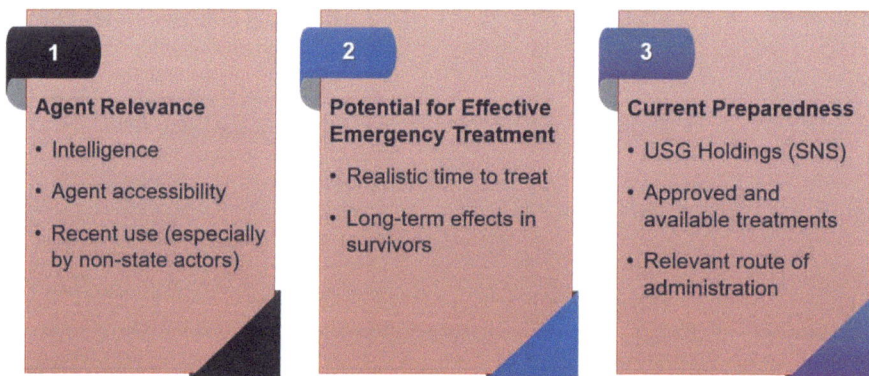

1 Agent Relevance
- Intelligence
- Agent accessibility
- Recent use (especially by non-state actors)

2 Potential for Effective Emergency Treatment
- Realistic time to treat
- Long-term effects in survivors

3 Current Preparedness
- USG Holdings (SNS)
- Approved and available treatments
- Relevant route of administration

FIGURE 6-4 BARDA's scheme for determining which chemical threats to prioritize for MCM development and production starts with an assessment by the IC of agent relevance followed by an evaluation of the potential for effective emergency treatment. The results of these assessments are used to identify and address gaps in current preparedness.
SOURCE: BARDA briefing to the committee.

have been developed by USAMRICD for a select number of chemicals, USG supports research to develop, improve, and optimize treatments for many as yet unaddressed chemical threats (Jett & Laney, 2001). CounterACT was created in addition to USAM-RICD due to significant differences between military and civilian scenarios including the demographics of the at-risk population. The U.S. Congress appropriated funds to the NIH to implement the National Strategic Plan and Research Agenda focused on understanding chemical toxicities and to use that knowledge to identify novel targets and develop promising candidate therapeutics.

CounterACT involves partnerships with other federal agencies, academia, and industry (see Table 6-2); its mission is to integrate cutting edge scientific research with the latest technological advances and medicine that could facilitate a rapid response during a chemical emergency. NIAID's current research priorities are based on DHS list of chemical agents (classified) and the Chemical and Biological Defense Program. NIAID is currently engaging with the broader academic research community to evaluate these priorities as a means to strengthen response based on identified toxidromes. The recommendations on the research priorities and the collaborations can be found in the 2003 NIAID Summary of the NIAID Expert Panel Review on Medical Chemical Defense Research (NIAID, 2003).

In the NIAID summary, the participants provided recommendations in the area of medical research for chemical defense that drove to the research objectives for this research program. CounterACT has provided outcomes that fulfill its objectives of stimulating and facilitating the development of a collaborative research community. These efforts include connecting research communities within academia, government (e.g., DoD and HHS), and industry partners.

In an information-gathering meeting with the committee, Dr. Yeung reported that with these collaborative research efforts, the NIH CCRP program has developed a pipeline of several MCMs (see Table 6-2).

TABLE 6-2 Drugs or Therapeutics Under the NIH CCRP Program and Collaborators

Name of Drug or Therapeutic	Function	Collaborator
Galantamine	neuroprotectant for organophosphate (OP) intoxication	Countervail Corporation
Midazolam	anti-seizure drug for OP intoxication	Pfizer
tPA	treatment for airway cast obstruction induced by inhalation of sulfur mustard	Genentech
R-107	treat inhalation chlorine injuries	Radical Therapeutics, Inc
TRPV4 Channel Blocker	treatment for inhalation chlorine injuries	GSK
Tezampanel	anti-seizure for benzodiazepines-resistant OP intoxication	PRONIRAS
Ganaxolone	anti-seizure/neuroprotectant for OP intoxication	Marinus Pharmaceuticals
INV-102	treats sulfur mustard-induced ocular injury	Invirsa and USAMRICD

The CounterACT program has further transitioned promising MCM candidates to BARDA. The CounterACT-BARDA facilitates partnerships with the pharmaceutical industry, which is essential for providing an integrated, systematic approach to the development of the necessary vaccines, drugs, therapies, and diagnostic tools for public health medical emergencies including chemical accidents, incidents, and attacks. Thus far, BARDA has obtained FDA approval for Argentum's Silverlon for sulfur mustard burns, Meridian's Seizalam for status epilepticus, and Primary Response Incident Scene Management (PRISM) Guidance for Mass Decontamination (U.S. DHHS, n.d.).

BARDA funds companies to drive innovation in using existing pharmaceutical products for new use in medical countermeasures (repurposing drugs) as well as developing broad-spectrum treatment measures (threat agnostic countermeasures). Figure 6-5 illustrates an integrated approach of coupling spectroscopy and computation to analyze unknown samples that could be chemical threats. For example, BARDA, in partnership with Johnson & Johnson, created an initiative called Blue Knight (Johnson and Johnson, 2022) which is dedicated to

> anticipating potential health security threats, activating the global innovation community, and amplifying scientific and technological advancements with the aim to prepare for and respond to our rapidly evolving global health environment.

Public Health Emergency Medical Countermeasures Multiyear Budget (PHEMCE MYB) for MCM development and stockpiling for HHS agencies; NIH, Assistant Sec-

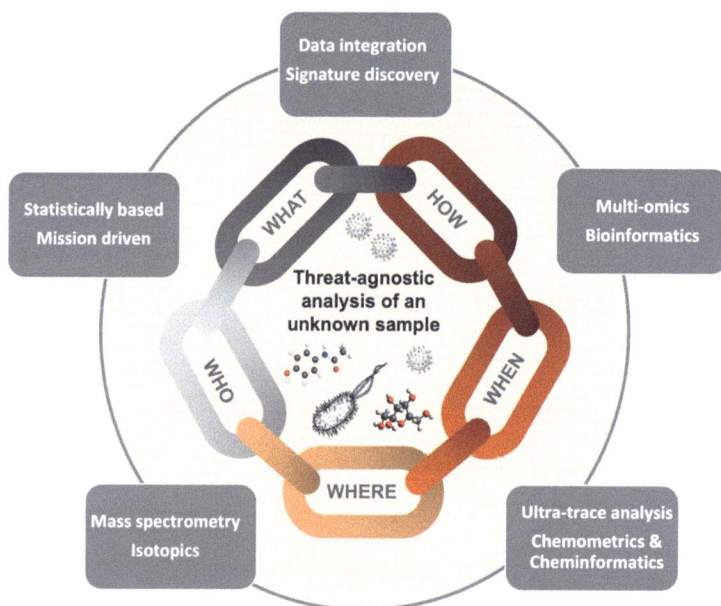

FIGURE 6-5 Actionable forensic attribution at the speed of relevance requires multidisciplinary approaches.
SOURCE: Pacific Northwest National Laboratory (PNNL).

retary for Preparedness Response (ASPR)-BARDA-, and ASPR-SNS, and FDA for the period between 2022–2026 (see Table 6-3). The PHEMCE MYB funding will be used to develop and support the transition of ten MCM candidates from BARDA's Project BioShield (PBS) to stockpiling by the SNS by fiscal year (FY) 2026. The chemical portfolio is allotted $1.5 billion over five years and its portfolio includes six NIH institutes and BARDA. The portfolio also includes funding to support the sustainment of SNS's current level of preparedness through replacement of expiring anticonvulsants, nerve agent antidotes, and other supportive medical materials (BARDA, 2023).

There are several challenges faced by researchers and drug companies in developing MCMs for use in a chemical emergency or terrorist attack since their characteristics, route, and time of administration have to be relevant for use in mass casualty scenarios (Jett and Laney, 2021).

Another challenge that BARDA faces is obtaining rapid availability of MCM supplies should there be a terrorist attack. Access to key components of the pharmaceutical industrial base and supply chain, which are currently manufactured outside the United States, is an issue. A majority of the Application Programming Interfaces (APIs) and their chemical compounds needed for critical medicines are also manufactured abroad. BARDA has created strategic partnerships with industry to expand pharmaceutical manufacturing in America with the aim to increase the domestic industrial base to allow for the additional raw material and consumables production necessary to support the manufacturing of therapeutics and vaccines during an emergency (BARDA, 2023).

The NIH, DTRA, and BARDA programs are using basic and translational research for MCM development and have broadened the base of collaborations among academic researchers and laboratories across the nation. Due to the ease of availability or access to toxic industrial chemicals/toxic industrial materials (TICs/TIMs), their use in terrorist activities has been enhanced. Therefore, the development of MCMs for next-generation chemical threats is needed (Casillas et al., 2021). Although NIH and BARDA have been successful in leveraging innovation and partnerships to accelerate the development of MCMs and obtain FDA licensure and clinical application for many, there is limited availability of FDA-approved MCMs for chemical threat exposures.

TABLE 6-3 Estimated Total PHEMCE Spending by HHS Division and Fiscal Year (dollar in millions)

Division	FY 2022	FY 2023	FY 2024	FY 2025	FY 2026	Total
NIH	$2,835	$2,825	$3,065	$3,131	$3,199	$15,055
ASPR BARDA	$1,818	$1,973	$13,192	$12,394	$10,928	$40,305
ASPR SNS	$845	$975	$1,963	$1,588	$1,439	$6,809
FDA	$216	$224	$371	$519	$527	$1,857
TOTAL	$5,714	$5,997	$18,590	$17,632	$16,093	$64,025

SOURCE: DHHS: Administration for Strategic Preparedness and Response, 2023.

6.2 SUMMARY

The current set of U.S. strategies, operational plans, and other resources has helped establish a network of capable first responder communities prepared for various chemical incidents, regardless of their cause. Tools aimed at improving communication for CBRN response have been discontinued, which could impede robust coordination across federal and SLTT organizations. Including input from first responders and export control communities will ensure timely threat assessments and protocol adjustments. This inclusion is especially important for strategies that involve whole of government sharing: U.S. Global Deterrence Framework. The committee found that the NRF adequately addressed chemical terrorism categorically under ESF #10 and Oil and Hazardous Materials Response. Nonetheless, translating U.S. strategies and frameworks into operational practice for chemical terrorism response remains a challenge. Given the complexity of the chemical threat landscape and the need for effective response coordination within the USG, the committee suggested current counterterrorism and emergency preparedness programs maintain a strong initiative of interagency exercises and trainings that focus on enhancing communication and resource sharing.

REFERENCES

BARDA. 2023. Medical Countermeasures Capacity Partnerships. https://www.medicalcounter measures.gov/barda/influenza-and-emerging-infectious-diseases/coronavirus/pharma ceutical-manufacturing-in-america.

Casillas, R., N. T-Singh, J. Gray. 2021. "Special Issue: Emerging Chemical Terrorism Threats." *Toxicology Mechanisms and Methods* 31(4): 239–241. https://doi.org/10.1080/15376516.2 021.1904472.

ChemResponder Steering Committee. 2020. "ChemResponder Network Who We Serve." https://www.fema.gov/sites/default/files/2020-07/fema_cbrn_chemresponder_fact-sheet.pdf.

CISA. 2018. SAFECOM Nationwide Survey, 2018. https://www.cisa.gov/safecom/safecom-nationwide-survey.

DHHS (U.S. Department of Health and Human Services). n.d. https://aspr.hhs.gov/AboutASPR/ProgramOffices/BARDA/Pages/default.aspx.

DHHS, Administration for Strategic Preparedness and Response. 2023. Public Health Emergency Medical Countermeasures Enterprise: Multiyear Budget: Fiscal Years 2022–2026.

DHS (U.S. Department of Homeland Security). 2021. Chemical Security Analysis Center, Jack Rabbit II. https://www.dhs.gov/sites/default/files/publications/2021_0205_csac-jack_rabbit_ii_factsheet_2-5-2021_pr_508.pdf.

DHS. 2006. Communicating in a Crisis: Chemical Attack. https://www.dhs.gov/sites/default/files/publications/prep_chemical_fact_sheet.pdf.

DOT (U.S. Department of Transportation). 2023. Pipeline and Hazardous Materials Safety Administration, Office of Hazardous Material Safety, 10 Year Incident Summary Reports. https://www.phmsa.dot.gov/hazmat-program-management-data-and-statistics/data-operations/incident-statistics.

ECPC (Emergency Communications Preparedness Center). 2021. Annual Strategic Assessment. https://www.cisa.gov/sites/default/files/cisa/21_0702_2019%20Emergency%20Communications%20Preparedness%20Center%20(ECPC)%20Annual%20Strategic%20Assessment%20Report_508c.pdf.

EPA (Environmental Protection Agency). 2008. ESF #10—Oil and Hazardous Materials Response Annex. https://www.fema.gov/pdf/emergency/nrf/nrf-esf-10.pdf.

FEMA (Federal Emergency Management Agency). n.d. ChemResponder Network. https://www.fema.gov/sites/default/files/2020-07/fema_cbrn_chemresponder_fact-sheet.pdf.

FEMA. n.d. Office of Emerging Threats. https://www.fema.gov/about/offices/response-recovery/emerging-threats.

FEMA. 2010. Developing and Maintaining Emergency Operations Plans, Comprehensive Preparedness Guide (CPG) 101 Version 2.0. November 2010.

FEMA. 2021. Oil and Chemical Incident Annex. (http://www.fema.gov/sites/default/files/documents/fema_incident-annex-oil-chemical.pdf.

Fox, S., and S. McMasters. 2022. Jack Rabbit III Initiatives: Chemical Threat Characterization through Experimentation for Strengthening Safety and Security of Critical Infrastructure. https://www.cisa.gov/sites/default/files/publications/2020-seminars-jack-rabbit-III-508.pdf.

Jett, D., and J. Laney. 2021. "Civilian Research on Chemical Medical Countermeasures.," *Toxicology Mechanisms, and Methods* 31(4): 242–243. https://doi.org/10.1080/15376516.2019.1669250.

Joint Counterterrorism Assessment Team. n.d. "First Responder Toolbox." Www.dni.gov. https://www.dni.gov/index.php/nctc-how-we-work/joint-ct-assessment-team/first-responder-toolbox.

Johnson and Johnson. 2022. Blue Knight. https://jnjinnovation.com/jlabs/blue-knight.

NIAID (National Institute of Allergy and Infectious Diseases). 2003. Summary of the NIAID Expert Panel Review on Medical Chemical Defense Research. https://www.niaid.nih.gov/sites/default/files/chem_report.pdf.

NIH/NLM (National Institutes of Health/National Library of Medicine. 2023. Wireless Information System for Emergency Responders (WISER). https://www.nlm.nih.gov/wiser/index.html.

NSS. 2022. https://www.whitehouse.gov/wp-content/uploads/2022/10/Biden-Harris-Administrations-National-Security-Strategy-10.2022.pdf,.

Savage, T. 2022. FBI, National Security Branch, briefing to the Chem Threats Committee. August 11, 2022.

START (National Consortium for the Study of Terrorism and Responses to Terrorism). 2022. Global Terrorism Database 1970–2020 [data file]. https://www.start.umd.edu/gtd.

U.S. Government Accountability Office. 2023. Chemical Weapons: Status of Forensic Technologies and Challenges to Source Attribution (GAO-23-105439). https://www.gao.gov/products/gao-23-105439.

White House. 2018. National Strategy for Countering Weapons of Mass Destruction Terrorism. https://www.hsdl.org/?view&did=819382.

7

Chemical Terrorism in the Era of Great Power Competition: Cross-Cutting Findings, Conclusions, Recommendations

---

Summary of Key Findings, Conclusions, and Recommendations

FINDING 7-1: The highest-level strategies of the United States, the National Security Strategy (NSC) and National Defense Strategy (NDS), have overtly shifted away from focusing on the threats from violent extremist organizations to great power competition (GPC) in recent years. This change indicates a shift in relative perceived threat and consequent prioritization and will impact efforts against chemical terrorism. Changes in strategy lead to changes in funding priorities, and while operational changes are anticipated from this major strategic shift, neither the mechanism, magnitude, nor timing are currently understood.

RECOMMENDATION 7-1: The shift in the global threat landscape has led to a corresponding shift in countering weapons of mass destruction (WMD) to a focus on GPC, but care should be taken to ensure existing capabilities focused on countering terrorism are maintained. Recommendations based on revised risk assessments that are aligned with new national-level priorities should be developed.

FINDING 7-2: The Department of Homeland Security (DHS), while acknowledging the national strategic shift to great power competition in the 2021 China Strategic Action Plan, has not published a strategy that both acknowledges the shift and also addresses chemical terrorism.

RECOMMENDATION 7-2: The DHS should develop strategies, including an updated chemical defense strategy, that consider the implications of the strategic shift to GPC, including potential resourcing shifts, on reduc-

continued

Summary Continued

ing the risk of chemical threats and chemical terrorism. Such strategies, whether public or not, should lead to specific, actionable plans and detail expected outcomes for counterterrorism activities in the context of current national strategic priorities. The committee acknowledges that such documents may be in progress.

FINDING 7-3: The U.S. Department of Defense (DoD) NDS acknowledges that terrorism risks may rise as program priorities shift to other priorities and other circumstances evolve.

RECOMMENDATION 7-3: The DoD should monitor risks associated with the shift in strategic focus and adapt if evidence of terrorist activities ramps back up.

RECOMMENDATION 7-4: The intelligence community and its offices throughout the departments with significant chemical terrorism roles and responsibilities (DoD, DHS, DOJ) should take steps to ensure that counter chemical weapons programs, whether state-based or by nonstate actors, are not technologically deterministic. This will require efforts to address gaps in knowledge or approaches which may arise as new personnel are hired as well as other transitions. The best way to do this needs to be determined by individual offices and agencies in consultation with the wider homeland security or defense community.

CONCLUSION 7-5: The shift in strategic focus to GPC will likely lead to reduced resources for countering weapons of mass destruction terrorism broadly, although the mechanisms, magnitude, and timing of those changes are currently poorly understood. How these changes are made is important; sudden changes without thoughtful preservation of functions could impede tactical readiness against chemical terrorist threats and increase risk in unforeseen or undesirable ways.

RECOMMENDATION 7-5: The DoD should conduct risk and threat assessments to understand how best to direct resources to address risks of chemical terrorism events in an era of GPC-focused strategies.

FINDING 7-6: The legislation establishing the Chemical Facility Anti-Terrorism Standards (CFATS) program (6 CFR Part 27) expired at the end of July 2023. Reauthorization will provide regulatory certainty for one of America's critical infrastructures in support of reducing the threat of chemical terrorism.

RECOMMENDATION 7-6: Congress should immediately reauthorize the CFATS program and consider long-term reauthorization.

FINDING 7-7: Current broadly extensible strategies could support effective identification, prevention, and response to the widest range of anticipated and yet-to-be-recognized chemical agents.

Summary Continued

CONCLUSION 7-7: Further adoption of approaches with broad extensibility can partially mitigate loss of focus on chemical terrorism due to the shift to GPC.

RECOMMENDATION 7-7: Federal agencies should prioritize broadly applicable approaches beyond the specific mission sets represented by the U.S. Army Research and Development Center for Chemical and Biologic Defense Technology (DEVCOM CBC), Biomedical Advanced Research and Development Authority (BARDA), and Cybersecurity and Infrastructure Security Agency (CISA), to all areas of the Countering Weapons of Mass Destruction and Terrorism (CWMDT) enterprise to maximize the United States' government capacity for appropriate response on time scales of relevance.

CONCLUSION 7-8: Strategy documents that include implementation plans with descriptions of current levels of inter- and intra-agencies coordination will significantly enhance communications across relevant entities. The areas of identify, prevent, counter, and response to chemical threats and chemical terrorism will especially benefit from this improvement. With respect to chemical terrorism events, communication between state and local law enforcement (LE) during an emergency could be impeded by classification issues.

RECOMMENDATION 7-9: WMD budgets should be aligned with evolving strategic priorities.

RECOMMENDATION 7-10: CWMDT budgets should incentivize activities to transition from promising research to operations.

FINDING 7-11: The material reviewed by the committee showed insufficient detail to allow a robust assessment of budgets likely to be required to implement strategies effectively, particularly for offices whose missions cover both chemical and biological threats.

CONCLUSION 7-11: Revised risk assessments are needed to reprioritize risks guided by recently issued strategies, so that strategy-aligned budgets can be created. To ensure a balance among efforts initiated by revised assessments, a distinction between countering chemical and countering biological efforts is needed.

The committee was tasked with evaluating strategies against chemical terrorism at a time of evolving national strategy. The United States' highest-level strategies recently explicitly shifted away from focusing primarily on violent extremist organizations (VEOs) to focusing on GPC,[1] which is indicative of a shift in perceived threat prioriti-

---

[1] Beginning with the 2018 *National Defense Strategy of the United States of America: Sharpening the American Military's Competitive Edge*, December, 2017, and continuing and increasing in the current administration's strategies.

zation. This shift in strategy does not mean terrorism is now regarded as unimportant; rather, it simply means that other issues have risen in importance, a fact which is most apparent in the National Security Strategy (NSS, 2022) and the 2022 National Defense Strategy (NDS, 2022). The Biden Administration NSS states:

> *The most pressing strategic challenge facing our vision is from powers that layer authoritarian governance with a revisionist foreign policy… a challenge to international peace and stability.* (Pg. 8)

The document does not ignore terrorism completely. The NSS also states:

> *America remains steadfast in protecting our country and our people and facilities overseas from the full spectrum of terrorism threats that we face in the twenty-first century.* (Pg. 30)

The NDS delineates four top-level defense priorities—none of which directly mention terrorism of any kind. Although terrorism, including chemical attacks and VEOs, are mentioned in the document, they are implicitly subordinated in the 2022 NDS by neither including them in the top-level priorities, nor mentioning them in the executive summary or the conclusion of the strategy.

Both of the above strategies portray a similar theme: While the threat of terrorism remains real, our nation is shifting its focus to prioritize different strategic threats, namely GPC over VEOs. With the apparent shift of strategic focus to GPC comes a shift in risk perception, risk assessment, and risk acceptance. Eventually, new strategies and risk assessments generate new mitigation strategies; new guidance, policies, and laws; and ultimately, new tactics, techniques, and protocols (TTPs) on the ground. The full ramifications of recent strategic shifts have not yet been realized at all levels of the government, nor at all agencies.

FINDING 7-1: The highest-level strategies of the United States, the NSS and NDS, have overtly shifted away from focusing on the threats from VEOs to GPC in recent years. This change indicates a shift in relative perceived threat and consequent prioritization and will impact efforts against chemical terrorism. Changes in strategy lead to changes in funding priorities, and while operational changes are anticipated from this major strategic shift, neither their mechanism, magnitude, nor timing is currently understood.

RECOMMENDATION 7-1: The shift in the global threat landscape has led to a corresponding shift in countering WMD to a focus on GPC, but care should be taken to ensure that existing capabilities focused on countering terrorism are maintained. Recommendations based on revised risk assessments that are aligned with new national-level priorities should be developed.

7.1 DEPARTMENT OF HOMELAND SECURITY (DHS) STRATEGY

DHS was created in response to the most significant international terrorism attack perpetrated against the United States. Because DHS has a primarily domestic security function, its risk and threat assessments will not necessarily follow the same pattern as DoD or the intelligence community IC. Furthermore, how the strategic shift from VEOs to GPC will impact DHS' strategic posture, programs, human resources, and missions is yet to be fully understood. Reducing the terrorism threat, including chemical terrorism, will continue to remain at the organization's core, as articulated in the overview of the DHS document, *Preventing Terrorism*, "Protecting the American people from terrorist threats is the reason DHS was created, and remains our highest priority" (DHS, n.d.). Preventing terrorism is one of 13 issues (with some overlapping scope) handled by DHS. Furthermore, DHS's 2020–2024 Strategic Plan outlines six strategic goals that align with general national prosperity under VEO-focused or GPC-focused strategies, but they do not specifically acknowledge GPC as a top national threat (see Figure 7-1).

DHS overtly acknowledges the shift to GPC in its 2021 China Strategic Action Plan (SAP)—which predates the most recent NSS—and which asserts that its fundamental mission of safeguarding the homeland, upholding DHS's values, and preserving the American way of life remains, even in the evolving geopolitical environment (DHS, 2021). The SAP addresses the following areas: maritime security, cybersecurity and critical infrastructure, trade and economic security, and border security and immigration. Aspects of terrorism including chemical terrorism are absent from the discussion. The only chemical threat specifically mentioned was China's direct and indirect involvement in supplying fentanyl and its precursors to drug cartels and transnational criminal organizations, contributing to more than 70,000 deaths in the United States in 2019 (DHS, 2021, 6, 9).

More recently, in April 2023, DHS published the Third Quadrennial Homeland Security Review (QHSR) (DHS, 2023). While recognizing the NSS, the QHSR only implicitly emphasizes the shift in national strategy to GPC by highlighting issues facing the United States as a result of GPC. The QHSR does not explicitly discuss China's role in the Strategic Competition section of the document but does detail actions to address issues facing the U.S. because of strategic competition with China specifically. The document embraces the role of increased partnerships in DHS's strategy, a theme found in the NSS and NDS. With respect to terrorism, the opening letter from Secretary Mayorkas states,

> *Today, the most significant terrorist threat stems from lone offenders and small groups of individuals, especially domestic violent extremists, while the threat of international terrorism remains as foreign terrorist organizations have proven adaptable and resilient over the past two decades and individuals inspired by their ideologies have continued to launch attacks in their names.*

More specific to chemical terrorism is the DHS Chemical Defense Strategy of December 2019, which the committee evaluated in detail. As of June 2023, the DHS Chemical Defense strategy has not been updated after the release of the DHS 20–24 strategic plan nor since national strategies have shifted their focus to GPC (DHS, 2022). QHSR does not specifically address chemical terrorism apart from other types of terrorism.

Strategic Goals	Strategic Objectives		CBP	CISA	FEMA	ICE	TSA	USCIS	USCG	USSS	HQ/Support
Goal 1: Counter Terrorism and Homeland Threats	1.1	Collect, Analyze, and Share Actionable Intelligence									
	1.2	Detect and Disrupt Threats									
	1.3	Protect Designated Leadership, Events, and Soft Targets									
	1.4	Counter WMDs and Emerging Threats									
Goal 2: Secure U.S. Borders and Sovereignty	2.1	Secure and Manage Air, Land, and Maritime Borders	■			■			■		
	2.2	Extend the Reach of U.S. Border Security	■			■					
	2.3	Enforce U.S. Immigration Laws	■			■					
	2.4	Administer Immigration Benefits						■			
Goal 3: Secure Cyberspace and Critical Infrastructure	3.1	Secure Federal Civilian Networks									
	3.2	Strengthen the Security and Resilience of Critical Infrastructure									
	3.3	Assess and Counter Evolving Cybersecurity Risks									
	3.4	Combat Cybercrime									
Goal 4: Preserve and Uphold the Nation's Prosperity and Economic Security	4.1	Enforce U.S. Trade Laws and Facilitate Lawful Trade and Travel	■								
	4.2	Safeguard the U.S. Transportation System	■				■				
	4.3	Maintain U.S. Waterways and Maritime Resources							■		
	4.4	Safeguard U.S. Financial Systems									
Goal 5: Strengthen Preparedness and Resilience	5.1	Build a National Culture of Preparedness									
	5.2	Respond During Incidents									
	5.3	Support Outcome-Driven Community Recovery									
	5.4	Train and Exercise First Responders									
Goal 6: Champion the DHS Workforce and Strengthen the Department	6.1	Strengthen Departmental Governance and Management	All DHS Headquarters Offices and Component counterparts								
	6.2	Develop and Maintain a High Performing Workforce									
	6.3	Optimize Support to Mission Operations									

FIGURE 7-1 Strategic goals of the U.S. DHS and the associated objectives. Colored boxes illustrate the key mission groups in the columns that are responsible for upholding the objects and goals: U.S. Customs and Border Protection (CPB), CISA, Federal Emergency Management Agency (FEMA), Immigration and Customs Enforcement (ICE), Transportation Security Agency (TSA), United States Citizenship and Immigration Services (USCIS), United States Secret Service (USSS), and Headquarters or Support (HQ/Support).
SOURCE: DHS, 2020.

In sum, DHS has released several documents outlining the organization's plans to address aspects of the shift to GPC or chemical terrorism. However, a key question remains: With a shift to GPC-focused strategies by the nation, are DHS strategies against chemical terrorism threats (and terrorism threats more broadly) appropriately prioritized and resourced?

FINDING 7-2: The DHS, while acknowledging the national strategic shift to great power competition in the 2021 China SAP, has not published a strategy that both acknowledges the shift and also addresses chemical terrorism.

RECOMMENDATION 7-2: The DHS should develop strategies, including an updated chemical defense strategy, that consider the implications of the strategic shift to GPC—including potential resourcing shifts—on reducing the risk of chemical threats and chemical terrorism. Such strategies, whether public or not, should lead to specific, actionable plans and detail expected outcomes for counterterrorism activities, in the context of current national strategic priorities. The committee acknowledges that such documents may be in progress.

7.2 DEPARTMENT OF DEFENSE STRATEGY

The shift to GPC also impacts the DoD, though differently than the domestically focused DHS. DoD's intersection with chemical terrorism is part of a broader concern about terrorism threats against U.S. assets—and those of our allies—overseas and about terrorist assets that might mature into a threat against the homeland. In the NDS, DoD embraces the shift to prioritizing GPC, which will likely lead to a reallocation of resources supporting the new prioritization. The NDS (2022) states:

This strategy will not be successful if we fail to resource its major initiatives or fail to make the hard choices to align available resources with the strategy's level of ambition." The NDS also states *"No strategy will perfectly anticipate the threats we may face, and we will doubtless confront challenges in execution.*

As noted at the beginning of this chapter, the shift in focus will likely lead to shifts in resources, which will inevitably affect risk profiles in other areas. The NDS is clear-eyed about this reality, which likely will help ensure that the department effectively implements the strategy and assesses its impact over time (DoD, 2022).

FINDING 7-3: The U.S. DoD NDS acknowledges that terrorism risks may rise as program priorities shift to other priorities and other circumstances evolve.

RECOMMENDATION 7-3: The U.S. DoD should monitor risks associated with the shift in strategic focus and adapt if evidence of terrorist activities ramps back up.

Although the NSS and NDS have been updated to a GPC focus, as of June 2023, joint doctrine reflected in Joint Publications (JP 3-11, JP 3-40, JP 3-41) has not been updated to reflect this shift since the release of the most recent NDS.[2]

---

[2] NOTE: The committee recognizes that at the time of producing the report, the newest NDS for CWMD may not be publicly released.

.. states that the:

Department of Defense derives its national strategic direction primarily from the President's guidance in the NSS, presidential directives, and other national strategic documents... (Pg. viii)

However, as doctrinal documents, the Joint Publications describe principles as to how the Joint Force fights wars that do not change with changes in strategy. As such, Joint Publications do not generally undergo revision and updating when shifts in strategy occur, unless the principles described in doctrine no longer apply or have changed in some way. Therefore, it remains to be seen if the Joint Publications will need to be revised specifically in response to the strategic shift to GPC.

7.3 INTELLIGENCE COMMUNITY STRATEGY

The committee also heard from key counterterrorism program managers from the IC in a public information-gathering meeting. One example of attitudes regarding counterterrorism during the shift to GPC was presented by Tom Breske, Senior Advisor for the Weapons of Mass Destruction Counterterrorism Team under the Directorate of Strategic Operational Planning at the National Counterterrorism Center (NCTC). Breske was posed the question:

What is the most under-recognized or un-recognized threat or problem (current or future) related to responding to and reducing chemical terrorism threats?

Breske responded,

As our Nation's focus shifts to the threat posed by nation-state near-peer competitors such as China, Russia, Iran, and North Korea, we [the United States] must maintain a watchful eye on nonstate actors and violent extremist organizations for indications and warnings of interest in accessing, procuring, manufacturing, training, and potential use of WMD, both at home and abroad. Understanding that an effective Counter WMD-Terrorism strategy depends upon an effective Counterterrorism capability, shifts in resources and priorities away from counterterrorism will likely have downstream impacts on our ability to counter chemical terrorism.

Breske's answer above speaks to the balancing act of changing strategic priorities when the United States does not (publicly) know the origin of the next WMD threat. This comment was similar to the responses of other U.S. officials briefing the committee. The shift to GPC is recognized and prioritized in strategy in the IC, but how it is operationalized with respect to chemical terrorism across the IC as a whole is not readily apparent.

RECOMMENDATION 7-4: The IC and its offices throughout the departments with significant chemical terrorism roles and responsibilities (DoD, DHS, DOJ) should take steps to ensure that counterchemical weapons programs, whether

state-based or by nonstate actors, are not technologically deterministic. This will require efforts to ensure as gaps in knowledge, approaches that may arise as new personnel are hired, and others transition. The best way to do this needs to be determined by individual offices and agencies in consultation with the wider homeland security or defense community.

7.4 CHEMICAL TERRORISM RISKS

In addition to the strategic shift to GPC, the NSS speaks with a sense of urgency in implementation. Without recapitulating the entire strategy, the urgency is best summarized in the final words of the NSS, "There is no time to waste" (NSS, 2022, Pg. 48).

With respect to chemical terrorism, none of the briefings to the committee (see List of Briefers in Appendix A) fully acknowledged the GPC as the top strategic priority of the United States nor the urgency desired in the most recent national strategies. To be fair, often the briefers focused on more operational aspects related directly to chemical defense and this was not in the original set of questions put forward by the committee. The committee acknowledges that the nation is in a dynamic state of developing new strategies.

If GPC intensifies, there are potential implications for chemical terrorism threats, beyond a possible reduction of resources available to address the threats. As discussed in Chapter 4, decisions states make may wittingly or unwittingly lead to a dramatic increase in the sophistication of chemical terrorism, in terms of both the agent employed and/or the means by which it is delivered. Another increased risk of the shift to GPC is that assumptions about what chemical terrorism will look like will increasingly be influenced and modeled on state-based programs, motivations, and thinking. While there certainly are lessons to be learned, the capacity, capability, and willingness to innovate—as well as network structures—differs significantly between states and nonstate actors. As attention on nonstate actors and threats decrease, the tendency to treat terrorism as a "lesser included" case or type of chemical weapons use may increase given constraints on budgets and time. Terrorism is different—a fact that is particularly important in the context of strategies to identify and counter, including deterrence.

CONCLUSION 7-5: The shift in strategic focus to GPC will likely lead to reduced resources for CWMDT broadly, although the mechanisms, magnitude, and timing are currently poorly understood. How these changes are made is important; sudden changes without thoughtful preservation of functions could impede tactical readiness against chemical terrorist threats and increase risk in unforeseen or undesirable ways.

RECOMMENDATION 7-5: The DoD should conduct risk and threat assessments to understand how best to direct resources to address risks of chemical terrorism events in an era of GPC-focused strategies.

7.5 APPROACHES TO IDENTIFY, PREVENT, COUNTER, AND RESPOND WITH BROAD APPLICABILITY

The following section describes key themes related to the use of broadly extensible strategies to identify, prevent, and respond as discussed across the previous chapters. For example, the surveillance-like use of zebra fish by DEVCOM CBC could also be applied to screening for toxicity when identifying chemical agents. Some agencies already perform counterterrorism activities and strategies that are similarly broad in their applicability, and which will be important to retain. A selection of these broad approaches is summarized and depicted in Figure 7-2.

Identify

"Identify" strategies must facilitate the discovery of the actor, intended agent(s), and delivery mechanisms and tactics. In terms of identifying an actor and intended delivery mechanisms and tactics, the FBI's current practice of informing export control and transportation security officials of new Tactics, Techniques, and Protocols (TTPs) utilized by specific VEOs facilitates more timely identification of VEO activities (FEMA, 2016). Meanwhile, in terms of identifying the agent being pursued by a VEO, being able to detect the fact that a chemical or class of chemicals that is somehow associated with a VEO is linked to a particular toxidrome can help identify which chemicals should be considered threats (e.g., surveillance-like application of DEVCOM CBC's work with zebra fish).

Prevent

As with "identify," making export control and transportation security officials aware of new TTPs utilized by a specific VEO can empower them to also effectively

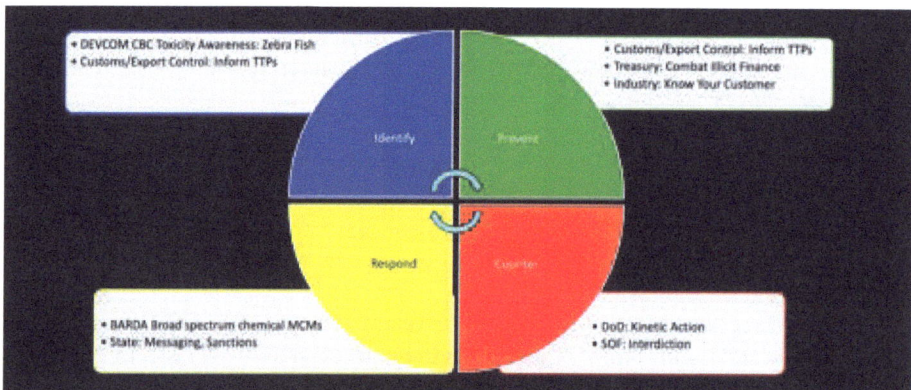

FIGURE 7-2 Conceptual framework of threat agnostic approaches to identify, prevent, counter, and respond.
SOURCE: Kabrena Rodda, 2023.

prevent a planned attack. Similarly, efforts undertaken to implement the 2022 National Strategy for Combatting Terrorist and Other Illicit Financing help to identify VEO activities which could in turn help prevent a planned attack (Treasury, 2023). Information derived from lines of effort in this strategy can support and inform Know Your Customer (KYC) initiatives undertaken by chemical producers and equipment manufacturers (SWIFT, n.d.).

At the time of writing this report, the committee learned that CFATS's statutory authorization was allowed to expire. Therefore, CISA

> *cannot enforce compliance with the CFATS regulations at this time. This means that CISA will not require facilities to report their chemicals of interest or submit any information in the Chemical Security Assessment Tool (CSAT, perform inspections, or provide CFATS compliance assistance, amongst other activities. CISA can no longer require facilities to implement their CFATS Site Security Plan or CFATS Alternative Security Program" (CISA, 2023).*

FINDING 7-6: The legislation establishing the CFATS program (6 CFR Part 27) expired at the end of July 2023. Reauthorization would provide regulatory certainty for one of America's critical infrastructures in support of reducing the threat of chemical terrorism.

Given the role CFATS played in preventing and countering chemical terrorism (see Chapter 5 full discussion), the committee directs the following recommendation to Congress. The American Chemistry Council (ACC) also supports the reauthorization of the CFATS program (ACC, 2023).

RECOMMENDATION 7-6: Congress should immediately reauthorize the CFATS program and consider long-term reauthorization.

Counter

Activities undertaken under "counter" largely fall under DoD or Special Operations Forces (SOF) and FBI and are perhaps the most broadly applicable to each CWMDT focus area. In the case of DoD, direct kinetic action against a VEO may be the most effective at countering a WMDT attack that is already underway. SOF could be made aware of specific supply nodes to facilitate such activities.

Respond

BARDA's practices of using existing chemical medical countermeasures (MCMs) for prioritized toxidromes and to promote decontamination as a first step whenever possible are likely to facilitate time-relevant response in the aftermath of an attack and to make sure first responders are aware of, and equipped to treat, the most likely toxidromes. More broadly, diplomatic efforts and messaging directed toward attributing attacks to specific VEOs and entities who support them can be highly effective without having to identify the specific chemical used in an attack (2018 National Strategy for Countering WMDT Terrorism; Joint Countering Weapons of Mass Destruction).

7.6 THREAT-AGNOSTIC APPROACHES TO MCMS AGAINST CHEMICAL THREATS

If resources for counterterrorism decrease due to the shift toward GPC, then a burden will be placed on existing programs to use their resources more efficiently in countering chemical threats. Despite the potential loss of focus on chemical terrorism, the growing trend toward more broadly extensible strategies being implemented by many agencies (see Chapter 2) may help reduce risk. By focusing response, particularly MCM, on the main physical and symptomatic effects of a chemical rather than the particular chemical or how and why it was used, responders may address the impact of chemical events more quickly.

Explicit adoption of a threat agnostic, or agent agnostic, approach in the context of response, (e.g., MCM development, and the explanation of the reasoning behind that change in approach) can also be found in the DoD's CBDP *Approach for Research, Development, and Acquisition of Medical Countermeasure and Test Products* (CBDP, 2022).

> *To better prepare the Joint Force against future and unknown threats, including naturally occurring emerging pathogens, the Chemical and Biological Defense Program (CBDP) will pivot away from viewing the threat landscape as a defined list of known biological and chemical agents toward removing or reducing the impact of agents' effects. This shift demonstrates how the CBDP will view medical countermeasure (MCM) response as a spectrum that requires investing in the development of broad-spectrum (or nonspecific) MCM and test products and establishing capabilities to rapidly develop narrow spectrum (or specific) MCM and test products. (Pg. 1)*

Notably, a broad spectrum, rather than specific one-bug(agent)-one-drug, approach has been a goal in the realm of biological threats for many years. In 2007, the CBDP initiated an effort that became known as the Transformational Medical Technologies Initiative (TMTI) to "develop broad-spectrum medical countermeasures against advanced bio-terror threats, including genetically engineered, intracellular bacterial pathogens and hemorrhagic fevers." (TMTI, 2007, Pg. 3)

TMTI identified: "the possibility that future state or nonstate adversaries could develop and deploy new genetically engineered biological threats for which current countermeasures would be ineffective and the time needed to develop defense would be insufficient" (TMTI, 2007, Pg. 3) as a driver for the major policy endeavor as part of science and technology efforts to respond to national security threats that the military might have to face in future years. In so doing, even as the threat evolves (e.g., use of different chemical agents or different types or sources of attack), a given MCM should continue to be effective.

Building on earlier work from the DoD's CBDP, Figure 7-3 shows BARDA's prioritization of five diagnostic toxidromes for chemical exposures (neurologic, pulmonary, respiratory, metabolic, vesicating) bypasses the need to identify the specific agent that

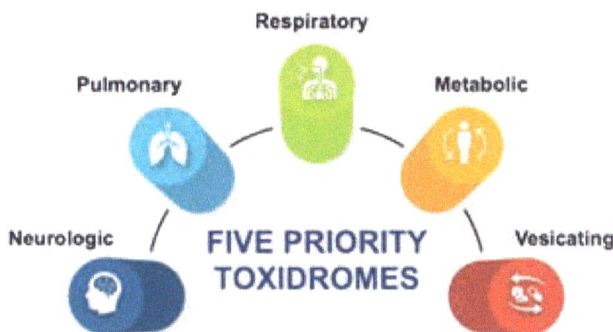

FIGURE 7-3 Five toxidromes are currently prioritized by BARDA's Chemical MCMs Unit.
SOURCE: BARDA Chemical MCMs Program.

caused the injury and allows for "broad spectrum therapeutic utility" (BARDA) using FDA approved drugs for treatment. The adoption of a toxidrome-based approach can be observed in BARDA's CBRN MCM chemical threat portfolio, which as of June 2023, lists 16 drugs undergoing different phases of development whose purpose is to treat both chemical-specific injury and non-agent-specific symptoms (BARDA, n.d.). Alteplase, for example, is a therapeutic that, if approved, will be the first drug to treat sulfur mustard inhalation or pulmonary exposure. RWJ-800088 thrombopoietin mimetic is another drug that aims to address both the radiation/nuclear and chemical threat areas by protecting vulnerable human cells from radiation and chemical exposures. It also accelerates recovery in the lungs and thrombocytopenia (low platelet counts in the blood). BARDA's toxidrome adoption and growing CBRN MCM portfolio have positioned the agency to more readily develop and deploy effective chemical medical countermeasures across multiple sectors to "treat the injury, not the agent" (BARDA, n.d.).

The committee endorses this approach to MCMs and notes that the same kind of conceptual approach may apply to identifying and preventing threats. Many counterterrorism programs have long taken this kind of approach. For example, major terrorist plots require financing, secrecy, and communications. As mentioned earlier, several intelligence, LE, and threat reduction programs seek to track and disrupt illicit financing, make it harder to operate without observation, and monitor or disrupt communications among threat actors. In an environment of increasingly constrained resources, extending this conceptual approach could be an efficient strategy.

> FINDING 7-7: Current broadly extensible strategies could support effective identification, prevention, and response to the widest range of anticipated and yet-to-be-recognized chemical agents.

CONCLUSION 7-7: Further adoption of approaches with broad extensibility can partially mitigate the loss of focus on chemical terrorism due to the shift to GPC.

RECOMMENDATION 7-7: Federal agencies should prioritize broadly appli-cable approaches beyond the specific mission sets represented by the U.S. Army DEVCOM CBC, BARDA, and CISA, to all areas of the CWMDT enterprise to maximize the United States' government capacity for appropriate response on time scales of relevance.

CONCLUSION 7-8: Strategy documents that include implementation plans with descrip-tions of current levels of inter- and intra-agencies coordination will significantly enhance communications across relevant entities. The areas of identify, prevent, counter, and response to chemical threats and chemical terrorism will especially benefit from this improvement. With respect to chemical terrorism events, communication between state and local law enforcement during an emergency could be impeded by classification issues.

7.7 SIMILARITIES AND CROSSOVER IN EFFORTS TO COUNTER THREATS FROM BIOTERRORISM AND CHEMICAL TERRORISM

Addressing biological and chemical threats requires inter- and multidisciplinary approaches that bridge the life, data, medical, physical, and social sciences, along with engineering, skill sets, and expertise. This includes the importance of data integration when information is coming in from different areas. Historically, much analysis has been technologically deterministic and/or based on limited empirical case studies and often limited analysis. While the United States tends to focus on technological solu-tions to both chemical terrorism and bioterrorism threats, there is less policy focus on motivation for both forms of terrorism.

A number of important and effective programs were developed after 9/11 to enhance the capabilities of the United States to prevent, prepare for, detect, respond, mitigate, and recover from terrorism events. Over time, many programs have been hampered by the gradual erosion of federal support and competing challenges. These include criti-cal diagnostic networks and resources, surveillance efforts (national and international), and exercise programs to integrate cross-sector emergency personnel groups together.

Coordination across and inside government agencies is important. While there is a good deal of coordination—some very effective—within the chemical terrorism defense communities and within the bioterrorism defense communities, crossover between the two communities is often lacking. The United States government (USG) would benefit from increased sustained interaction. Coordination across and among federal, county, state, local, territorial, and tribal (SLTT) governments is critical and often a challenge. Training and exercises are critical for planning, preparedness, response, and remediation.

Mis- and disinformation represent an expanding challenge for chemical and biological threats and require additional study to identify mechanisms to effectively counter their influence. These studies should inform U.S. policies to counter mis- and disinformation. The critical need for clear consistent information is recognized at the federal level in responding to chemical and bioterrorism incidents; how to deal with mis- and/or disinformation is often not included. (Perhaps a consequence of broader USG challenges of dealing with mis/disinformation rather than unique to any WMD). Uncertain information exists regarding interactions between the federal and state levels or at state levels.

- The approach to assessing and addressing biological and chemical threats should take human, plant, animal, and ecosystem health into account.
- In terms of attribution, reference collections and databases have been significantly improved or outright created for both chemical and biological threats. In the case of databases relating to bioterrorism threats, the level of curation (i.e., how good is the data), has not been pursued/validated as well as with chemical terrorism threats. Much of this may be due to the newness of the underlying technology (e.g., genetic and microbiome databases). This is a potential vulnerability that may not be well recognized (i.e., are we fooling ourselves into believing we have greater capabilities than we do?).
- New technologies are neither panaceas nor inherently threats. The underlying drivers and manner in which people may choose to use/misuse them are under-studied/under-recognized in U.S. policies.
- Experts do not agree on the nature, scope, or scale of biological and chemical terrorism threats. Lots of uncertainty surrounds this area.

7.8 BUDGET RECOMMENDATIONS

Through a series of information-gathering meetings, several briefers noted to the committee that budgets were inadequate to address the breadth of possible chemical threats. (See list of briefings in Appendix A.) Presenters also indicated concern that constrained budgets may make it difficult to invest sufficiently in promising low technology readiness level (TRL) concepts to enable them to mature sufficiently for transition into operational use. Even when countering WMD terrorism was among the highest priorities, some federal agencies had tightly constrained budgets to counter chemical terrorism. For example, in an information-gathering meeting, the committee was informed that the FBI's WMD directorate has had static staffing levels for the past 17 years. Aligning to the national strategic shift to GPC may compromise efforts to counter chemical terrorism in this agency. Regardless, the national strategies have evolved significantly; new budgets should follow suit.

RECOMMENDATION 7-9: WMD budgets should be aligned with evolving strategic priorities.

Strategies employing flexible funding, such as the State Department's Nonproliferation and Disarmament Fund (NDF), enable prompt, effective responses that build on activities across the USG and in cooperation with the Organization for the Prohibition of Chemical Weapons (OPCW) and allies.

NDF was created to enable the USG to respond quickly to "high-priority non-proliferation and disarmament opportunities." Many USG programs have geographic, budgetary, or political restrictions on their authorities. As explained on the NDF website:

> NDF funds are "no-year" (funds need not be expended in the fiscal year in which they are appropriated) to permit maximum flexibility in project execution and may be made available "notwithstanding any other provision of law."

NDF is a relatively small program (approximately $20 million per year vs. several hundred million dollars per year for cooperative threat reduction programs at the Departments of Defense and State) that has, because of its notwithstanding authority, supported an array of efforts from removal and destruction of the last of Libya's legacy chemical weapons program to removal of highly enriched uranium from a civilian facility in Serbia. Starting around 2017, there was an interagency agreement to apply these funds more flexibly to fill gaps where other programs lacked authority or funds to address threats before they became acute.

Despite the fact that many of the recommendations made in this report would not require a significant increase to the budget appropriation or authorization, the federal budget process presents significant barriers to effective transitioning of promising research and development (R&D). According to "U.S. Federal Scientific Research and Development: Budget Overview and Outlook," Federal funding for science and technology (S&T) is complicated by the U.S. budget process and the highly decentralized organization of federal R&D activities. The lack of a central mechanism to coordinate federal R&D programs leads to lengthy negotiations and decision-making among agencies, congressional committees (oversight and appropriation committees), and the White House. Beyond these challenges, various developments impacting the overall domestic and international strategic environment—such as the shift to GPC—can further complicate the effective transitioning of promising R&D. In this regard, the report states:

> Shifting priorities between presidential administrations, changes to the makeup and ideologies of Congress, and broader economic conditions in the United States at large have resulted in the inconsistent funding for R&D, especially for basic research, despite strong and consistent support from the American public (Evans et al., 2021).

The article also discusses the impact on S&T as a result of delayed approvals as well as abrupt changes in the budget. Delayed budget approvals severely disrupt agency operations as they must work under previous budget guidelines without knowing when or what the new fiscal budget will be. This uncertainty and possible budget disruption impacts the continuity of data collection, staffing, and the ability to start new projects and maintain large-scale facilities. An example of these challenges is evident in the Trump administration's de-emphasis of basic R&D in favor of development (efforts to take very mature R&D and transition it to operations) and then the Biden administration's reversal which proposed increases in all areas of R&D (basic, industrial, and development) as shown in Figure 7-4.

There are two high-level budget gaps among the strategies the committee evaluated. (1) Some of the newer U.S. strategies have shifted to a GPC-focus (NSS, NDS) and others have to a lesser extent (DHS). Resourcing differently focused strategies in a

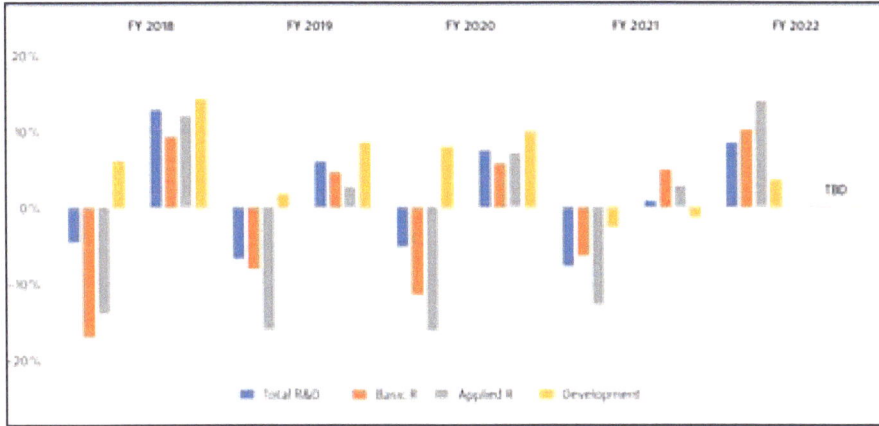

FIGURE 7-4 While President Trump called for reductions to federal basic and applied research funding (light) for fiscal year (FY) 2018–2021, congressional appropriations funded R&D agencies and activities more generously (dark). President Biden's first budget proposal requested increases to R&D for the upcoming FY 2022. FY 2021 data does not include COVID-19 relief appropriations.
SOURCE: AAAS, 2021.

hierarchy must be rationalized. (2) Detailed resourcing of new GPC-focused strategies has not yet occurred. The importance of aligning resourcing with strategy has been declared (NDS, 2022), but the rebalancing has not appeared in public budgets to date. As stated throughout this report, robust anti-terrorism efforts are still important but other efforts associated with GPC now have higher strategic priority.

> **RECOMMENDATION 7-10: Counter weapons of mass destruction (CWMDT) budgets should incentivize activities to transition promising research to operations.**

As stated earlier in this chapter, chemical terrorism and broader terrorism are deemphasized but not ignored in the shift to GPC. The highest-level strategies (NSS, NDS) speak to both an urgency to adopt GPC as a priority and a willingness to accept risk from other threats when doing so. In this dynamic context of a strategic shift with at present undefined budgetary implications, it is difficult for the committee to make specific budget recommendations in dollar amounts. Instead, the committee recommends that chemical terrorism risk assessments (e.g., full risks, threats only, national-level, state-level, and others) be performed in the context of the latest strategies to align budget priorities with strategic priorities and most clearly understand where and why the United States is accepting risk. Table 7-1 shows the budget functions and resources the committee believes should be considered under budgetary constraints that may result from the national strategic shift to GPC. These factors include risk priorities that are expressed in budget requests.

TABLE 7-1 Recommended Budget Priorities Based on National Strategic Shift to Great Power Competition

Budget Function or Resource	Benefit of Retention
Fund comprehensive risk assessments based on the priorities set forth in recent national security strategies.	Allows forward-thinking strategic planning and preparedness. Enables agility to focus on new priorities when national strategy evolves. Identifies alignment between funding emphasis and strategy. Identifies where risk is being accepted when alternate, more strategy-aligned, investments are made.
Maintain the IC capabilities and expertise specific to terrorist groups (VEOs and racially, ethically, and motivated violent extremists) and understand their motivations.	Ensures subject matter expertise in the terrorism threat space is retained. Allows for rapid identification of and adaptation to emerging threats.
Support basic scientific and social science research specifically related to countering chemical terrorism, (e.g., understanding social behavior related to emerging threats).	Retains a strong talent base to address future, perhaps unanticipated, chemical threats/substances and the motivations to use them. Threats change and without natural and social scientific research, it will be difficult to adapt to changes, or in some cases even understand that and/or why they have occurred.
Strengthen insider threat programs related to physical, cyberphysical, and cybersecurity across the chemical industry.	Secures physical facilities from being subverted to cause toxic releases or the theft of precursor chemicals. Protects vulnerable information systems from being used in espionage and chemical attacks.
Support training and exercises to advance international chemical security priorities through continued initiatives with, for example, the OPCW, Proliferation Security Initiative (PSI) partners, North Atlantic Treaty Association (NATO) allies, nongovernmental organizations, and other international stakeholders.	Increases capacity and tactical readiness internationally thereby decreasing global threat and decreasing reliance on U.S. assets to respond.
Fund initiatives that work with international partners to enhance chemical security and identify, prevent/counter, and respond to chemical threats worldwide.	Strengthens alliances and builds stronger communication networks among relevant international agencies.
Continue emphasizing programs employing threat-agnostic approaches to identify and respond to chemical attacks.	Enables more economical, efficient, and effective responses, especially in times when chemical terrorism, or other national security concerns, may be deemphasized.
Encourage more flexible capability portfolio management models and processes that reduce bureaucratic constraints to accelerate the adoption of emerging technologies. Utilize innovation like the cross-functional team program management approaches model.	Enables the flexibility to most promptly address evolving threats and to more effectively facilitate innovation adoption and integration. (Esper and Lee James, 2023)

FINDING 7-11: The material reviewed by the committee showed insufficient detail to allow a robust assessment of budgets likely to be required to implement strategies effectively, particularly for offices whose missions cover both chemical and biological threats.

CONCLUSION 7-11: *Revised risk assessments are needed to reprioritize risks guided by recently issued strategies so that strategy-aligned budgets can be created. To ensure a balance among efforts initiated by revised assessments, a distinction between countering-chemical and countering-biological efforts is needed.*

7.9 SUMMARY

The shift in the highest-level strategies of the United States from focusing on VEOs to GPC will impact efforts against chemical terrorism. While changes in funding priorities and operational adjustments are anticipated, the specific mechanism, magnitude, and timing are currently less understood. For example, sudden changes without thoughtful preservation of functions may hinder tactical readiness against chemical terrorist threats. Federal agencies should prioritize broadly applicable approaches to all areas of the CWMDT enterprise to maximize the USG capacity for appropriate response. Strategy documents that included detailed implementation plans for inter- and intra-agency coordination enhanced communication across relevant areas of identification, prevention, countering, and responding to chemical threats. However, communication challenges may still arise between state and local LE during emergencies due to classification issues related to chemical terrorism events.

The committee also cautioned the IC and relevant departments (DHS, DoD, DOJ) to avoid supporting counterchemical weapons programs that are technologically deterministic. Instead, the committee highlighted the need to address gaps in knowledge and approaches, such as involving social scientists with regional perspectives. A recommendation that DHS develop an updated chemical defense strategy that considers the implications of the GPC shift on reducing the risk of chemical threats and terrorism was made. Similarly, the committee observed that DoD's counterterrorism programs would adapt better to the shifting national-level priorities if the department closely monitors terrorism risks. In practice, DoD would conduct risk and threat assessments to understand how best to direct the limited resources. Through their assessment, the committee emphasized that the revised risk assessments are needed to reprioritize risks and create CWMDT strategy-aligned budgets. These budgets should incentivize activities to transition promising research to operations. A distinction between countering-chemical and countering-biological efforts is necessary to balance different efforts resulting from risk assessments. Regardless of any agency defining toxins as falling under chemical or biological, toxins should be included as a threat. Finally, the committee emphasizes the importance of ensuring the continuity of authorization, in particular CFATS, abilities of critical programs, which are essential in safeguarding the nation from chemical terrorism activities.

REFERENCES

AAAS (American Association for the Advancement of Sciences). 2021. Interactive Dashboards. https://www.aaas.org/programs/r-d-budget-and-policy/interactive-dashboards-0.

ACC (American Chemistry Council). 2023. ACC Statement Regarding Expiration of Essential Chemical Security Program Key to Combating Terrorism. https://www.americanchemistry. com/chemistry-in-america/news-trends/press-release/2023/acc-statement-regarding-expiration-of-essential-chemical-security-program-key-to-combating-terrorism

BARDA (Biomedical Advanced Research and Development Authority). n.d. Medical Counter-measures Count — They Work to Safeguard Us. https://medicalcountermeasures.gov/barda/cbrn#portfolio.

CBDP (Chemical and Biological Defense Program). 2022. Approach for Research, Development, and Acquisition of Medical Countermeasures and Test Products. https://media.defense.gov/2023/Jan/10/2003142624/-1/-1/0/approach-rda-mcm-test-products.pdf.

CISA (Cybersecurity and Infrastructure Security Agency). 2023. Chemical Facility Anti-Terrorism Standards. https://www.cisa.gov/resources-tools/programs/chemical-facility-anti-terrorism-standards-cfats.

Department of State. Office of the Nonproliferation and Disarmament Fund. N.d. https://www.state.gov/bureaus-offices/under-secretary-for-arms-control-and-international-security-affairs/bureau-of-international-security-and-nonproliferation/office-of-the-nonproliferation-and-disarmament-fund/

DHS (Department of Homeland Security). n.d. Preventing Terrorism Overview. https://www.dhs.gov/preventing-terrorism-overview.

DHS. 2021. The DHS Strategic Plan to Counter the Threat Posed by The People's Republic of China. https://www.dhs.gov/sites/default/files/publications/21_0112_plcy_dhs-china-sap.pdf.

DHS. 2022. Chemical, Biological, Radiological, and Nuclear Defense Research, Development, Test, and Evaluation Strategy: Fiscal Years 2021-2025. https://www.dhs.gov/sites/default/files/2022-03/22_0316_st_cbrn_rdt%26e_strategy_fy21-25.pdf.

DHS. 2023. The Third Quadrennial Homeland Security Review. https://www.dhs.gov/sites/default/files/2023-04/23_0420_plcy_2023-qhsr.pdf.

DOD. 2022. National Defense Strategy. https://www.defense.gov/National-Defense-Strategy/Risk Management Section.

DOD. 2007, Transformational Medical Technologies Initiative (TMTI). Fiscal Year 2007 (FY 2007) Congressional Report. https://biosecurity.fas.org/resource/documents/dod_2007_transformational_medical_technologies_initiative.pdf

Esper, M., and D. Lee J. 2023. Want More Pentagon Innovation? Try This Experiment.

Evans, K. M., K. R. W. Matthews, G. Hazan, and S. Kamepalli. 2021. U.S. Federal Scientific Research and Development: Budget Overview and Outlook. Rice University Baker Institute for Public Policy. https://www.bakerinstitute.org/research/us-federal-scientific-research-and-development-budget-overview-and-outlook.

FEMA (Federal Emergency Management Agency). 2016. National Prevention Framework. Retrieved from https://www.fema.gov/sites/default/files/2020-04/National_Prevention_Framework2nd-june2016.pdf.

NSS. 2022. https://www.whitehouse.gov/wp-content/uploads/2022/10/Biden-Harris-Administrations-National-Security-Strategy-10.2022.pdf.

SWIFT (Society for Worldwide Interbank Financial Telecommunication). n.d. Know Your Customer. https://www.swift.com/your-needs/financial-crime-cyber-security/know-your-customer-kyc.

Treasury (U.S. Department of Treasury). 2022. National Strategy for Combating Terrorist and Other Illicit Financing. https://home.treasury.gov/system/files/136/2022-National-Strategy-for-Combating-Terrorist-and-Other-Illicit-Financing.pdf.

Appendix A

U.S. Government Strategies and Other Documents Considered

TABLE A-1 U.S. Government Strategies and Other Documents Considered

Document	Year	Overview (Taken from Document/Source)
WHITE HOUSE/INTERAGENCY		
Interim National Security Strategic Guidance	2021	Presents President Biden's vision for how America will engage with the world among shifting global dynamics. Departments and agencies are directed to align their actions with this guidance, even as they begin work on a National Security Strategy (NSS). There is no explicit mention of chemical weapons or chemical terrorism.
National Security Strategy	2017	Organized around four pillars of effort, the first of which is "Protect the American People, the Homeland, and the American Way of Life." A goal within this pillar is to defend against weapons of mass destruction (WMD) threats, and priority actions are identified.
National Strategy for Counterterrorism of the United States of America	2018	Describes the United States government (USG) approach to countering nonstate WMD threats, emphasizing the need for continuous pressure against WMD-capable terrorist groups, enhanced security for dangerous materials throughout the world, and increased burden sharing among our foreign partners. Outlines five strategic objectives that emphasize prevention, deterrence, detection, identification, and response.

continued

TABLE A-1 Continued

Document	Year	Overview (Taken from Document/Source)
Presidential Policy Directive (PPD)-8: National Preparedness	2011	Aims at strengthening the security and resilience of the United States through systematic preparation for the threats that pose the greatest risk to the security of the nation, including acts of terrorism, cyberattacks, pandemics, and catastrophic natural disasters. Under this directive, the Secretary of Homeland Security is responsible for coordinating domestic all-hazards preparedness efforts of federal departments and agencies in consultation with other levels of government, nongovernmental organizations, private sector partners, and the public.
PPD-21: Critical Infrastructure Security and Resilience	2013	Outlines a national effort to strengthen and maintain secure, functioning, and resilient critical infrastructure. The chemical industry is designated as the first of 16 national critical infrastructure sectors.
Homeland Security Presidential Directive (HSPD)-4: National Strategy to Combat Weapons of Mass Destruction	2002	Outlines three pillars of effort to counter the threat of WMDs: counterproliferation to combat WMD use; strengthened nonproliferation to combat WMD proliferation; and consequence management to respond to WMD use. The Strategy also details four "cross-cutting enabling functions" to be pursued: intelligence collection and analysis on WMD, delivery systems, and related technologies; research and development to improve our ability to respond to evolving threats; bilateral and multilateral cooperation; and targeted strategies against hostile states and terrorists.
HSPD-5: Management of Domestic Incidents	2003	Establishes a single, comprehensive national incident management system to enhance the ability of the United States to manage domestic incidents. This directive gives the Secretary of Homeland Security responsibility for managing domestic incidents, including incidents related to chemical terrorism.
HSPD-9: Defense of U.S. Agriculture and Food	2004	Establishes a national policy to defend the agriculture and food system against terrorist attacks, major disasters, and other emergencies.
HSPD-18: Medical Countermeasures Against Weapons of Mass Destruction	2007	Describes the principles from which national guidance is derived for addressing the challenges presented by the diverse chemical, biological, radiological, and nuclear (CBRN) threat spectrum, optimizing the investments necessary for medical countermeasures development, and ensuring that USG activities significantly enhance domestic and international response and recovery capabilities. Outlines the chemical threats for which the development of targeted medical countermeasures might be warranted.
National Strategy for CBRNE Standards	2011	Describes the need for chemical, biological, radiological, nuclear, and high yield explosives (CBRNE) standards. The Strategy specifies high-level goals, identifies lead activities to accomplish these goals, and provides the foundation to bridge current gaps. It establishes a structure to facilitate the coordination of CBRNE investments and activities among agency leaders, program managers, the research and testing community, and the private sector.

TABLE A-1 Continued

Document	Year	Overview (Taken from Document/Source)
DEPARTMENT OF DEFENSE		
U.S. Army Medical Research and Development Command (USAMRDC) Handbook	2023	Establishes how each direct reporting agency unit works both internally and within the command structure to deliver emerging science and cutting-edge material to U.S. soldiers. Reporting agencies include the Chemical Biological Radiation and Nuclear Defense Research Coordinating Office Chemical Biological Defense Program (CBDP) (https://mrdc.amedd.army.mil/assets/docs/media/USAMRDC-Handbook.pdf)
DoD Strategy for Countering Weapons of Mass Destruction (CWMD)	2014	Represents the DoD's response to the WMD threat. Specifies desired end states, prescribes priority objectives, delineates a strategic approach for achieving those objectives, and outlines the countering WMD activities and tasks necessary for success. Presents four priority objectives to define a comprehensive response to the WMD challenge and focus on shaping the environment, cooperating with partners, and prioritizing early action.
DoD Directive (DoDD) 5160.05E: Roles and Responsibilities Associated with the Chemical and Biological Defense Program	2017, rev. 2019	Establishes policy and assigns responsibilities associated with the CBDP research, development, and acquisition of CBRD capabilities required to support CWMD missions as set forth in the DoD Strategy for CWMD and DoDD 2060.02. Designates and defines the role of the Secretary of the Army as the DoD Executive Agent for the CBDP.
DoD Instruction (DoDI) 3020.52: DoD Installation Chemical, Biological, Radiological, Nuclear, and High-Yield Explosive (CBRNE) Preparedness Standards	2012	Implements policy, assigns responsibilities, and prescribes procedures to establish and implement a program for a global DoD installation hazard response to manage the consequences of a CBRNE incident. It provides guidance for the establishment of a CBRNE preparedness program for emergency responders at all DoD installations.
DoDI 2000.21: DoD Support to International Chemical, Biological, Radiological, and Nuclear (CBRN) Incidents	2016, rev. 2017	Establishes policy and assigns responsibilities for DoD support to the USG response to international CBRN incidents. Policy states that DoD will conduct international CBRN-response operations to protect U.S. citizens, deter the use of WMD, minimize hazards and effects of CBRN incidents, and alleviate effects of such incidents.
Chairman of the Joint Chiefs of Staff Instruction (CJCSI) 3125.01D: Defense Response to Chemical, Biological, Radiological, and Nuclear (CBRN) Incidents in the Homeland	2015, current as of 202	Provides the CJCS policy guidance and operational instructions for DoD response to CBRN incidents in the homeland.
CJCSI 3214.01E: Defense Support for Chemical, Biological, Radiological, and Nuclear Incidents on Foreign Territory	1 2015, current as of 2021	Provides guidance for assistance provided by U.S. military resources in support of USG objectives to prepare for and respond to CBRN incidents that occur on or impact foreign territory.

continued

TABLE A-1 Continued

Document	Year	Overview (Taken from Document/Source)
Description of the National Military Strategy	2018	Unclassified summary of the Strategy which provides the Joint Force a framework for protecting and advancing U.S. national interests. The Strategy implements the policy and strategy direction provided in the 2017 National Security Strategy, the 2018 National Defense Strategy, the Defense Planning Guidance, and other documents. Articulates a continuum of strategic direction to frame global integration into three strategy horizons to meet the challenges of the existing and future security environment.
National Military Strategy to Combat Weapons of Mass Destruction	2006	Defines a strategic end state, military strategic objectives, and the missions and means to achieve them. Provides a framework to the DoD on which to base deliberate planning, coordination activities, operations, and capabilities development.
Chemical and Biological Defense Program (CBDP) Strategic Plan	2008	Guides the actions of the CBDP and outlines the strategic priorities to accomplish four overarching goals over the next 10 to 15 years. One such goal is: "Define and develop future capabilities to increase significantly our ability to dissuade, deter, defend against, and defeat any future adversary in any CBRN threat environment."
CBDP Annual Report to Congress, Public Summary	2021	Assesses and evaluates the DoD Fiscal Year 2020 chemical and biological defense efforts and overall readiness to fight and win in a chemically and biologically-contaminated environment.
CBDP Annual Report to Congress	2018	Focuses on the readiness of the DoD to respond to current and emerging threats and highlights important collaborations, research, and development activities to address novel threats. Highlighted are opportunities to strengthen the readiness of the Joint Force to operate in a contaminated environment.
Commander's Handbook for Strategic Communication and Communication Strategy	2009	Predoctrinal document on strategic communication (SC) and development of communication strategy at all levels of command. Provides fundamental principles and best practices as a bridge between current practice in the field and migration into doctrine. As such, it is a useful tool for identifying to improve/strengthen interagency communication and coordination in preventing, countering, and responding to WMDT involving chemical threats. https://apps.dtic.mil/sti/pdfs/ADA525371.pdf
DTRA-JSTO eBook	N/A	Outlines the mission, strategy, and capabilities of Defense Threat Reduction Agency Joint Science and Technology Office (DTRA JSTO). Describes the CBDP and DTRA JSTO's role in supporting disruptive scientific and technological advancements to protect the warfighter and the nation.

TABLE A-1 Continued

Document	Year	Overview (Taken from Document/Source)
DEPARTMENT OF HOMELAND SECURITY		
National Infrastructure Protection Plan (NIPP): Partnering for Critical Infrastructure Security and Resilience	2013	Guides the national effort to manage risk to the nation's critical infrastructure. Presents an integrated approach with partnerships among owners and operators; federal, state, local, tribal, and territorial (FSLTT) governments; regional entities; nonprofit organizations; and academia to manage the risks from significant threats and hazards to physical and cyber critical infrastructure. The approach addresses the need to identify, deter, detect, disrupt, and prepare for threats and hazards to the nation's critical infrastructure.
National Incident Management System (NIMS)	2017	NIMS guides all levels of government, nongovernmental organizations (NGOs), and the private sector to work together to prevent, protect against, mitigate, respond to, and recover from incidents. https://www.fema.gov/sites/default/files/2020-07/fema_nims_doctrine-2017.pdf
HHS Chemical Hazards Emergency Medical Management Response Guidance		Planning, medical response, and decontamination guidance for conventional chemicals and chemical warfare agents. Goals of the documents: Enable first responders, first receivers, other healthcare providers, and planners to plan for, respond to, recover from, and mitigate the effects of mass-casualty incidents involving chemicals; provide a comprehensive, user-friendly, web-based resource that is also downloadable in advance, so that it would be available during an event if the internet is not accessible; goals above are from the above top-level document found at https://chemm.hhs.gov/index.html.
		Similarly, 2015–2018 HHS/BARDA released the Primary Response Incident Scene Management (PRISM) Guidance documents. PRISM Volume 1: Strategic guidance relevant to senior incident commanders, https://www.medicalcountermeasures.gov/BARDA/Documents/PRISM%20Volume%201_Strategic%20Guidance%20Second%20Edition.pdf
		PRISM Volume 2: Reviews the processes involved in mass patient disrobe and decontamination, the rationale that underpins each process, and guidance for first responder training/exercising. https://www.medicalcountermeasures.gov/media/36873/prism-volume-2.pdf
		PRISM Volume 3: The tactical guidance mass patient disrobe and decontamination which aims to provide all federal, and STLL first responders with a simple, readily accessible guide to critical aspects of the incident response processes. https://www.medicalcountermeasures.gov/BARDA/Documents/PRISM%20Volume%203__Operational%20Guidance%20Second%20Edition.pdf CDC Chemical Emergency Guidance for the General Public. https://www.cdc.gov/chemicalemergencies/index.html

continued

TABLE A-1 Continued

Document	Year	Overview (Taken from Document/Source)
CLASSIFIED DOCUMENTS No Information		
National Security Presidential Memo/NSPM-36		
National Defense Strategy	2022 & 2023	
United States Global Campaign to Deter the Use of Chemical Weapons by State and Not-State Actors		

TABLE A-2 List of Organizations that Briefed the Committee

Federal Agency	Organization	Acronym	Briefer
United States House of Representatives	House Armed Services Committee		Shannon Green
DoD	Office of the Secretary of Defense (Policy)	OSD(P)	Robert Thompson
DoD	Defense Threat Reduction Agency, Joint Science and Technology Office	DTRA JSTO	Ronald K. Hann, Jr.
DoD	Joint Program Executive Office for Chemical, Biological, Radiological, and Nuclear Defense	JPEO-CBRND	Daniel J. McCormick
DoD	Office of the Assistant Secretary of Defense for Nuclear, Chemical, and Biological Defense Programs	OASD, NCB, or CBD	Ian Watson
DoD	U.S. Army Combat Capabilities Development Command Chemical Biological Center	CCDC CBC	Robert Kristovich and Joy Ginter
DoD	U.S. Army Medical Research Institute of Chemical Defense	USAMRICD	
DoD	20th Chemical, Biological, Radiological, Nuclear, Explosives Command	20th CBRNE Command	
Pentagon Briefings	Under Secretary of Defense for Policy, Principal Deputy Assistant Secretary of Defense	DASD	DASD Watson, IC Participants
DHS	Countering Weapons of Mass Destruction Office and Chemical Coordination Group	CWMD & CCG	Mark Kirk
DHS	Federal Emergency Management Agency	FEMA	Lito Ignacio

TABLE A-2 Continued

Federal Agency	Organization	Acronym	Briefer
DHS	Cybersecurity and Infrastructure Security Agency and Chemical Facility Anti-Terrorism Standards	CISA & CFATS	Annie Hunziker and Kelly Murray
DoD	Defense Threat Reduction Agency Cooperative Threat Reduction	DTRA, CTR	Pat Becker and Michelle Nalabandian Scott
DHS	Chemical Security Analysis Center	CSAC	Shannon Fox
FBI	Weapons of Mass Destruction Directorate	WMDD	Todd Savage
FBI	Chemical Biological Countermeasures Unit	CBCU	Scott Sharp
FBI	Intelligence Analysis Section		Mathew Hendley and Patrick McNellis
FBI	Laboratory Division		Doug Anders
NIH/ NIAID	Chemical Countermeasures Research Program	NIAID, CCRP	David Yeung
DHHS	Biomedical Advanced Research and Development Authority	BARDA	Judy Laney
NCTC	Weapons of Mass Destruction Counter Terrorism Group	WMD-CT	Thomas Breske
DoD	United States Special Operations Command	SOCOM	Ruth Berglin, Alissa Ackley, and Justin Gorkowski
State Department	The Bureau of International Security and Nonproliferation	ISN	Michael Wipper, Allison Tolbert, Costa Nicolaidis, and Kaitlyn Hudson

Appendix B

Acronym/Initialism List

ACC	American Chemistry Council
AI	Artificial Intelligence
APHIS	Animal and Plant Health Inspection Service (USDA)
API	Application Programming Interface
ASPR	Assistant Secretary for Preparedness and Response
ASTDR	Agency for Toxic Substances and Disease Registry
ATF	Bureau of Alcohol, Tobacco, Firearms and Explosives
BARDA	Biomedical Advanced Research and Development Authority
BWC	Biological Weapons Convention
CABNSAD	Chemical and Biological Weapons Non-State Adversary Database
CAMEO	Computer-Aided Management of Emergency Operations
CBC	Chemical Biological Center
CBCU	Chemical Biological Countermeasures Unit
CBDP	Chemical and Biological Defense Program (DoD)
CBRN	Chemical, biological, radiological, and nuclear
CBRNE	Chemical, biological, radiological, nuclear, and high yield explosives
CCDC	Collegiate Cyber Defense Competition
CCG	Chemical Coordination Group
CCMDs	Coordinating with Combatant Commands
CCRP	Chemical Countermeasures Research Program
CDC	Centers for Disease Control and Prevention (DHS)
CDER	Center for Drug Evaluation and Research (FDA)
CEPPO	Chemical Emergency Preparedness and Prevention Office (EPA)
CFATS	Chemical Facility Anti-Terrorism Standards

CFS	Cell-free synthesis
CHEMM	Chemical Hazards Emergency Medical Management
CISA	Cybersecurity and Infrastructure Security Agency (DHS)
CJCS	Chairman of the Joint Chiefs of Staff
CJCSI	Chairman of the Joint Chiefs of Staff Instruction
CM	Consequence management
CMCU	Consequence Management Coordination Unit
CNGB	Chief of the National Guard Bureau
COI	Chemicals of interest
CONUS	Continental USA
CounterACT	Medical chemical countermeasures against chemical threats
COVID	Coronavirus disease
CPB	Customs and Border Protection (DHS)
CRISPR	Clustered regularly interspaced palindromic repeats
CSAC	Chemical Security Analysis Center
CSB	Chemical Safety and Hazard Investigation Board
CT	Office of Counter Terrorism (DOS)
CTR	Cooperative Threat Reduction (DOS)
CW, CWA	Chemical weapon, chemical weapon agent
CWC	Chemical Weapons Convention
CWMD	Countering weapons of mass destruction/chemical WMD
CWMDT	Countering Weapons of Mass Destruction and Terrorism
DASD	Under Secretary of Defense for Policy, Principal Deputy Assistant Secretary of Defense
DEA	Drug Enforcement Administration
DevCom CBC	Army R&D Center for Chemical and Biologic Defense Technology
DHHS	Department of Health and Human Services
DHS	Department of Homeland Security
DOC	Department of Conservation
DoD	Department of Defense
DoDD	Department of Defense Directive
DoDI	DoD Instruction
DOE	Department of Energy
DOG	Digital Operations Guide
DOI	Digital object identifier
DOJ	Department of Justice
DOS	Department of State
DOT	Department of Transportation
DTRA	Defense Threat Reduction Agency (DoD)
DVE	Domestic violent extremist
ECBC	Edgewood Chemical Biological Center
ECC	Office of Export Control Cooperation (DOS)

ECPC	Emergency Communication Preparedness Center
ENMOD	Environmental Modification Techniques
EOP	Executive Office of the President
EPA	Environmental Protection Agency
EPC	Explosive precursor chemical
ESF	Emergency support function
EXBS	Export Control and Border Security Program (DOS)
FBI	Federal Bureau of Investigation
FDA	Food and Drug Administration
FEMA	Federal Emergency Management Administration
FOUO	For official use only
FY	Fiscal year
GAO	Government Accountability Office
GPC	Great power competition
GWOT	Global War on Terrorism
Hazmat	Hazardous materials
HIRA	Hazard identification and risk assessment
HQ/Support	Headquarters or Support (DHS)
HSPD	Homeland Security Presidential Directive
I&A	Office of Intelligence and Analysis (DHS)
IAFC	International Association of Fire Chiefs
IC	Intelligence community/Industrial chemicals
ICE	Immigration and Customs Enforcement (DHS)
ICS	Incident Command System
IED	Improvised explosive device
INTERPOL	International Criminal Police Organization
IS	Islamic State
ISIL	Islamic State of Iraq and the Levant
ISIS	Islamic State in Iraq and Syria
ISN	International Security and Nonproliferation Bureau (DOS)
IST	Inherently Safer Technology
JFCs	Joint Force Commanders
JP 3-40	Joint Publication, Countering Weapons of Mass Destruction, 2019
JP 3-41	Joint Publication, Chemical, Biological, Radiological, and Nuclear Response, 2016
JPEO	Joint Program Executive Office
JS	Joint Staff
JSTO	Joint Science and Technology Office

LE	Law enforcement
LLNL	Lawrence Livermore National Laboratory
MCF	Microbial cell factories
MCM	Medical countermeasure
ML	Machine learning
MNSA	Office of Multilateral Nuclear and Security Affairs
NASEM	National Academies of Sciences, Engineering, and Medicine
NATO	North Atlantic Treaty Organization
NCB	Nuclear, chemical, and biological
NCBA	Nationwide Communications Baseline Assessment
NCBC	National Counterproliferation Biosecurity Center
NCTC	National Counterterrorism Center
NDAA	National Defense Authorization Act
NDF	Nonproliferation and Disarmament Fund (DOS)
NDS	National Defense Strategy
NDU	National Defense University
NECP	National Emergency Communications Plan
NGO	Nongovernment organization
NIAID	National Institute of Allergy and Infectious Disease
NICC	National Infrastructure Coordinating Council
NIH	National Institutes of Health
NIMS	National Incident Management System
NIPP	National Infrastructure Protection Plan
NIU	National Intelligence University
NNSA	National Nuclear Security Administration (DOE)
NRF	National Response Framework
NRP	National Response Plan
NSAID	Nonsteroidal anti-inflammatory drug
NSC	National Security Council
NSM	National Security Memorandum
NSS	National Security Strategy
NTA	Nontherapy ancillary
O9A	Order of Nine Angles
OASD	Office of the Assistant Secretary of Defense
OBP	Office of Bombing Protection (DHS)
OCONUS	Outside the Contiguous United States
ODNI	Office of the Director of National Intelligence
OET	Office of Emerging Threats
OIP	Office of Infrastructure Protection (DHS)
OP	Organophosphate
OPCW	Organization for the Prohibition of Chemical Weapons

OSD(P)	Office of the Secretary of Defense (Policy)
OSHA	Occupational Safety and Health Administration
OSTP	Office of Science and Technology Policy
OUSD(P)	Office of the Undersecretary of Defense for Policy

PANTHER	Probabilistic Analysis of National Threats and Risk Program
PBA	Pharmaceutical-based agent
PBS	Project BioShield
PHEMCE MYB	Public Health Emergency Medical Countermeasures Multiyear Budget
PHMSA	Pipeline and Hazardous Materials Safety Administration (DOT)
POICN	Profiles of incidents involving CBRN and non-state actors
POW	Prisoner of war
PPD	Presidential Policy Directive
PPE	Personal protective equipment
PRC	People's Republic of China
PRISM	Primary Response Incident Scene Management
PSI	Proliferation Security Initiative

| QHSR | Quadrennial Homeland Security Review |
| QSAR | Quantum structure activity relationship |

R&D	Research and Development
RBPS	Risk-based Performance Standards
RCA	Riot-control Agent
RDT&E	Research, Development, Testing, and Evaluation
REMV	Racially and ethnically motivated violent extremism
RMP	Risk Management Program

S&T	Science and Technology
SAFECOM	Safer America Through Effective Public Safety Communication
SAP	Strategic Action Plan
SC	Strategic Communication
SLTT	State, local, territorial, and tribal
SME	Subject matter expert
SNRA	Strategic National Risk Assessment
SNS	Social Networking Service
SOF	Special Operations Forces
SOT	Statement of task
STEM	Science, technology, engineering, and mathematics

TCE	Threat Credibility Evaluation
THI	Toxic inhalation hazard
THIRA	Threat and Hazard Identification and Risk Assessment

TIC(s)	Toxic industrial chemical(s)
TIM(s)	Toxic industrial material(s)
TMTI	Transformational Medical Technologies Initiative
TRL	Technology Readiness Levels
TSA	Transportation Security Agency (DHS)
TTP	Tactics, techniques, and protocols

UNODC	United Nations Office on Drugs and Crime
UNSCR 1540	United Nations Security Council adopted resolution 1540
USAMRICD	U.S. Army Medical Research Institute of Chemical Defense
USCIS	United States Citizenship and Immigration Services (DHS)
USDA	Department of Agriculture
USG	United States government
USMS	United States Marshals Service
USSOCOM	United States Special Operations Command
USSS	United States Security Service (DHS)
USSTRATCOM	U.S. Strategic Command

VEO	Violent extremist organizations
VX	An extremely toxic chemical compound

WISER	Wireless Information System for Emergency Responders
WMD	Weapons of mass destruction
WMDD	Weapons of Mass Destruction Directorate
WMDSG	Weapons of Mass Destruction Strategic Group
WMDT	Weapons of mass destruction terrorism
WTC	World Trade Center
WWI, WWII	World War 1, World War 2

Appendix C

Committee Biographies

Timothy J. Shepodd (*Chair*) retired in 2023 from Sandia National Laboratories, where he was a senior manager, Mission Engineering Sciences. Dr. Shepodd was at Sandia for over 35 years working on various aspects of national security, including chemical munition destruction/neutralization, explosives chemistry, controlled substance synthesis, materials science for nuclear deterrence, and special communications. He participated in or chaired various committees chartered by the Program Executive Office for Assembled Chemical Weapons Alternatives while the final United States chemical weapons destruction plants were designed and built. Dr. Shepodd holds three R&D 100 awards and 38 patents, including one for the explosive destruction system used extensively to neutralize chemical munitions. He received a Ph.D. in chemistry from the California Institute of Technology and received his live chemical weapons training in the United Kingdom at the Porton Down facility.

Margaret E. Kosal (*Vice Chair*) is an associate professor in the Sam Nunn School of International Affairs at Georgia Institute of Technology. Formally trained as an experimental scientist, Dr. Kosal earned a Ph.D. in chemistry from the University of Illinois at Urbana-Champaign. Her research explores the relationships among technology, strategy, and governance. She focuses on two areas that often intersect: reducing the threat of weapons of mass destruction and understanding the geopolitics of emerging technologies. Dr. Kosal is the cofounder of a sensor company, where she led research and development of medical, biological, and chemical sensors, as well as explosives detection systems. Dr. Kosal currently serves as joint faculty appointee at Savannah River National Laboratory and previously has served as a senior advisor to the Chief of Staff of the U.S. Army, as science and technology advisor within the Office of the Secretary of Defense, and as an associate to the National Intelligence Council. She is the recipient of multiple awards including the Office of the Secretary of Defense Award for Excellence. Her numerous publications include *Nanotechnology for Chemical and Biological Defense*, the first and most rigorous to consider how nanotechnology may enable or be adapted for defensive purposes, along with potential misuse and proliferation risks. In 2017, she was appointed the editor-in-chief of the Cambridge University Press Journal *Politics and the Life Sciences*.

Gary A. Ackerman is an associate professor in the College of Emergency Preparedness, Homeland Security and Cybersecurity at the University at Albany, State University of New York. In addition, he is associate dean for Research in the College, and the founding director of the Center for Advanced Red Teaming. Previously, Dr. Ackerman held the posts of research director and then director of the Unconventional Weapons and Technology Division at the National Consortium for the Study of Terrorism and Responses to Terrorism. He has also served as the director of the Weapons of Mass Destruction Terrorism Research Program at the Center for Nonproliferation Studies in Monterey, California. Dr. Ackerman's research focuses on assessing emerging threats and understanding how terrorists and other adversaries make tactical, operational, and strategic decisions, especially how they innovate in their use of weapons and tactics. Much of his work in this area is centered on the motivations and capabilities for nonstate actors to acquire and use chemical, biological, radiological, and nuclear weapons. Dr. Ackerman is the coeditor of *Jihadists and Weapons of Mass Destruction* and author of over seventy publications. He has also testified on terrorist motivations for using nuclear weapons before the Senate Committee on Homeland Security and is a periodic consultant for the U.S. Government on counterterrorism issues. He served as a consultant to Emergent BioSolutions in 2018 and received compensation for these services. Dr. Ackerman completed his Ph.D. in war studies at King's College in London.

Philipp C. Bleek is a professor in the Nonproliferation and Terrorism Studies Program, faculty affiliate at both the James Martin Center for Nonproliferation Studies and Center on Terrorism, Extremism, and Counterterrorism, and coordinator of the Cyber Collaborative, all at the Middlebury Institute of International Studies at Monterey. He previously served as senior advisor to the Assistant Secretary of Defense for Nuclear, Chemical, and Biological Defense Programs in the U.S. Department of Defense. Dr. Bleek works on the causes, consequences, and amelioration of chemical, biological, radiological, and nuclear weapons threats, posed by both states and nonstate actors, at the intersection of academia, nongovernmental organizations, and government. Dr. Bleek has held fellowships at the Harvard Kennedy School's Belfer Center for Science and International Affairs, Council on Foreign Relations, Center for Strategic and International Studies, and Center for a New American Security, among others. In addition to his current faculty position, he has taught at Georgetown University and in the Department of Defense Senior Leader Development Program. He is a former term member of the Council on Foreign Relations and a fellow of the Truman National Security Project. Dr. Bleek holds a Ph.D. in government, with a concentration in international relations, from Georgetown University; a master of Public Policy, with a concentration in international policy and economic development, from the Harvard Kennedy School; and a B.A. in public and international affairs from Princeton University.

Gary S. Groenewold is recently retired from 35 years of service as a chemist at Idaho National Laboratory. During this span, Dr. Groenewold's broad scientific research interests involved investigations of reactivity and measurement of molecular and atomic species of importance in military, nuclear, and industrial enterprises. A significant fraction of his research involved the development of chemical measurement approaches for chemical warfare agents, which enabled improved insight into agent fate and transport in natural and industrial environments. He has authored approximately 150 peer-reviewed publications, many of which address chemical warfare agents. He is a member of the American Chemical Society and the American Society for Mass Spectrometry. Dr. Groenewold has coauthored six National Academies reports in the area of chemical demilitarization and served as the chair of the Chemical Demilitarization Committee. Dr. Groenewold holds a Ph.D. in chemistry from the University of Nebraska.

David J. Kaufman is the vice president and director for Safety and Security at CNA. He is responsible for executive management of CNA's work in the areas of public safety, homeland security, emergency management, and public health. From 2009–2015, Kaufman served as the associate administrator for Policy, Program Analysis, and International Affairs at the Federal Emergency Management Agency. Kaufman teaches in Georgetown University's graduate program in Emergency and Disaster Management; lectures for the Naval Postgraduate School's Center for Homeland Defense and Security; and has previously served as a committee chair and roundtable member for the National Academies of Science, Engineering and Medicine. Kaufman holds a Master of Public Policy degree from the University of Michigan; a B.A. in international relations, political science, and history from the University of Wisconsin-Madison; and is a graduate of the Center for Homeland Defense and Security's Executive Leaders Program.

Kabrena E. Rodda is a senior scientist at Pacific Northwest National Laboratory, providing strategic direction on research to improve U.S. capabilities against chemical threats. She is a retired U.S. Air Force (USAF) Colonel. During her USAF career, she managed a nonproliferation program and later advised on chemical issues at the National Counterproliferation Center. Dr. Rodda was a United Nations Special Commission inspector and laboratory chief in Iraq in 1995 and 1998 and provided consequence management advice for the 2000 Sydney Olympics. In 2012, she published a book-length policy paper against synthetic drugs titled *Legal Highs: U.S. Policy for the New Pandemic*. In 2017 and 2018, she led chemical threat response workshops at the Organisation for Prohibition of Chemical Weapons (OPCW) and headed the writing team for the American Chemical Society (ACS) policy statement, "Preventing the Reemergence of Chemical Weapons." Dr. Rodda is a recipient of the OPCW Director General's Medal, the Secretary of Energy Appreciation Award, and the Secretary of the USAF Research and Development Award. She is a member of ACS's International

Activities Committee, the American Academy of Forensic Science, and the International Society for the Study of Emerging Drugs. She holds a Ph.D. in forensic toxicology; three M.S. degrees in chemistry, project and systems management, and national security studies; and a B.S. in chemistry.

Neera Tewari-Singh is an assistant professor at the Department of Pharmacology and Toxicology at Michigan State University. With an extensive background in molecular biology and toxicology, her research focus is on developing medical countermeasures against chemical threats and environmental exposures that can cause harmful effects, chemical emergencies, and mass casualties. Dr. Tewari-Singh is a principal investigator in several National Institutes of Health Countermeasures Against Chemical Threats and the Department of Defense programs. She has published her work extensively and received numerous honors and awards, including the Society of Toxicology Association of Scientists of Indian Origin Young Investigator Award, Ocular Toxicology Innovation and Impact Award, and the Dermal Toxicology Best Paper of the Year award. Dr. Tewari-Singh received a Ph.D. from the Jawaharlal Nehru University, New Delhi.

Guy Valente (*member from 2/15/2022 to 1/6/2023*) oversees emergency preparedness activities and emergency medical services for the County of El Dorado in Northern California, where his attention is currently focused on COVID-19 and wildfire response efforts. Prior to assuming his current role in 2020, Valente served as an Inspector and Program Officer for the Organisation for the Prohibition of Chemical Weapons (OPCW), where he managed capacity-building programs for developing States Parties in the area of chemical emergency response. He also oversaw the destruction of chemical weapons stockpiles in China, Russia, and Libya, and had five deployments to Damascus under both the Fact-Finding Mission and a United Nations/OPCW Joint Mission in Syria. Valente is a former Hazardous Materials Handling technician, current paramedic, and an Advanced Hazardous Materials Life Support instructor. He holds a master's in public policy degree from the University of York in the U.K. and a B.A. in political science from the University of North Carolina.

Usha Wright is president of SHARE Africa, working in rural Kenya, as well as vice chair of the Board of Directors at Scenic Hudson, an environmental nongovernmental organization. Previously, she served as senior vice president, Environmental Safety and Health functions for ITT Corporation and before that, at Ciba Geigy (now Novartis), a pharmaceutical and agricultural manufacturer, and finally as general counsel and executive vice president at an environmental engineering corporation in New York. Wright has travelled extensively to manufacturing and research sites for the purpose of improving and mitigating environmental and safety risks of diverse operations, working with lawmakers as well as regulators. Her expertise is in chemical safety and related laws, in both domestic and international arenas. Wright previously served on a National Academies committee focused on the risks and minimizing the potential of diversion of research chemicals for terrorism purposes. She received a J.D. from Rutgers Law School.

Appendix D

Strategy Assessment Rubric

The following rubric was applied by the committee in their evaluation of the following areas: Identify, Prevent/Counter, and Response. The major categories examined were A, B, and C, which evaluate the existence of a genuine strategy, the sufficiency of a strategy to meet the chemical threat over a required time frame, and the feasibility of the strategy, respectively. The subcategories (A1–A3, B1 and B2, and C1–C3) consider different criteria as shown in the table.

Category	Subcategory	Criteria
A A genuine strategy exists	A1	(1) above exists [minimum 1 goal + definition of success].
	A2	For each goal in (1) above, there is at least one (2) above.
	A3	(1) and (2) are COHERENT (i.e., explicit and mutually consistent).
B The strategy is sufficient to meet the threat over the required timeframe of interest	B1	The goal(s) collectively encompass [identifying/preventing/countering/responding to/recovering from] the level and type of threat likely to emerge in the timeframe.
	B2	The policies, plans, and resource allocations are sufficient to achieve the goal(s) (or directly the level of threat likely to emerge in the timeframe).
C The strategy is feasible	C1	All the elements of the strategy, which are required to fulfill (B) above are also legally feasible.
	C2	All the elements of the strategy, which are required to fulfill (B) above are also fiscally feasible.
	C3	All the elements of the strategy, which are required to fulfill (B) above are also politically feasible.

Legend

1. = Well-defined goal(s), including a definition of "success."
2. = A set of policies, plans, and resource allocations designed to meet the corresponding goal(s) [minimum 1 plan/policy/resource allocation for each goal].

Steps to Take
1. Define or describe what "Success" looks like.
2. Use all evidence gathered to rate component A1.

Rating Scale A1	Criteria
Inadequate	There is *no evidence* that the United States possesses even one key goal aimed at the above categories.
Partially Inadequate	There is *some, but only partial*, evidence that the United States possesses at least one key goal aimed at the above categories of chemical terrorism.
Partially Adequate	There *is evidence* that the United States possesses at least one key goal aimed at the above categories, but *there is not a clear definition of success associated with it.*
Adequate	There *is evidence* that the United States possesses at least one key goal aimed at the above categories and *there is a clear definition of success.*

List all the identified goals of the strategy (goals are listed as *a, b, c, d,* etc.).

Rate Components A2a, A2b, A2c, etc.

Rating Scale A2a	Definition
Inadequate	There is *no evidence* that the United States has any policies, plans, or resource allocations to address Goal *a*.
Partially Adequate	There *is some*, but only partial, evidence that the United States has policies, plans, and/or resource allocations specifically designed to address Goal *a*.
Adequate	There *is evidence* that the United States has policies, plans, and/or resource allocations specifically designed to address Goal *a*.

Apply this exercise to the rest of the goals: A2b, A2c, A2d, etc.

Rate A2 Overall.

Rating Scale A2 Overall	Criteria
Inadequate	A2a, A2b, etc. are all "Inadequate." There are no policies, plans, or resource allocations for any of the strategy's goals.
Partially Inadequate	At least one of A2a, A2b, etc. is "Inadequate." At least one of the strategy's goals lacks corresponding policies, plans, and resource allocations (unless another strategy includes the same goal and is judged as "Adequate" above).
Partially Adequate	At least one of A2a, A2b, etc. is "Partially Adequate." There is only partial evidence that at least one of the strategy's goals possesses corresponding policies, plans and/or resource allocations (unless another strategy includes the same goal and is judged as "Adequate" above).
Adequate	All of A2a, A2b, etc. are "Adequate." All of the strategy's goals have policies, plans, and/or resource allocations specifically designed to address them.

Use all evidence to rate component A3.

Rating Scale: A3	Criteria
Inadequate	There are major internal contradictions between goals and/or policies in the strategy
Partially Inadequate	There are some internal contradictions between goals and/or policies, but these are relatively minor in the judgment of the committee.
Partially Adequate	All of the goals and policies, plans, and resource allocations in the strategy are internally consistent but not all have been explicitly documented.
Adequate	All of the goals and policies, plans, and resource allocations in the strategy are both explicitly documented and internally consistent with one another.

Combine components to rate Adequacy of "A "Overall.

Rating Scale: Combined A	Criteria
Inadequate	If any of A1, A2 or A3 is rated "Inadequate."
Partially Inadequate	If all of A1, A2, or A3 are rated at least as "Partially Inadequate," but not all are "Partially Adequate" or "Adequate."
Partially Adequate	If all of A1, A2, and A3 are rated at least as "Partially Adequate," but not all are "Adequate."
Adequate	If all of A1, A2, and A3 are rated as "Adequate."

Use all evidence rate component B1.

Rating Scale: B1	Criteria
Inadequate	The goal(s) collectively do not encompass one of the above category threats likely to emerge in the timeframe.
Partially Inadequate	The goal(s) collectively at least partially encompass the above category either the level or the type of threat likely to emerge in the timeframe, but not both.
Partially Adequate	The goal(s) collectively partially encompass the above category at the level and type of threat likely to emerge in the timeframe.
Adequate	The goal(s) collectively encompass the above category at least the level and type of threat likely to emerge in the timeframe.
Exceed	The goal(s) collectively encompass the above category at the level and type of threat likely to emerge in the timeframe, and even beyond this level and/or nature of threat.

Use all evidence to rate B2 for each goal, B2*a*, B2*c*, B2*d*, etc.

Rating Scale: B2a	Definition
Inadequate	Existing policies, plans, and resource allocations taken together *are insufficient to* achieve Goal *a* (and by extension the level of threat likely to emerge in the timeframe).
Partially Adequate	Existing policies, plans, and resource allocations taken together *are possibly, but not certainly,* sufficient to achieve Goal *a* (and by extension the level of threat likely to emerge in the timeframe).
Adequate	Existing policies, plans, and resource allocations taken together *are clearly sufficient* to achieve Goal *a* (and by extension the level of threat likely to emerge in the timeframe).
Exceed	Existing policies, plans, and resource allocations taken together *exceed what is necessary* to achieve Goal *a* (and by extension the level of threat likely to emerge in the timeframe).

Apply this exercise to the rest of the goals: B2b, B2c, B2d, etc.

Rate B2 Overall.

Rating Scale: B2 Overall	Criteria
Inadequate	B2a, B2b, etc. are all "Inadequate." There are no policies, plans, or resource allocations for any of the strategy's goals.
Partially Adequate	All of B2a, B2b, B2c, etc. are rated at least as "Partially Adequate" unless the goal(s) that are labeled "Partially Adequate" are covered by other equivalent goals in other strategy documents which themselves are rated "Adequate."
Adequate	All of B2a, B2b, B2c, etc. are rated at least as "Adequate."
Exceed	All of B2a, B2b, B2c, etc. are rated at least as "Adequate" and one or more or rated as "Exceed."

Combine components to rate Adequacy of "B "Overall.

Rating Scale: B Overall	Criteria
Inadequate	If any of B1 or B2 is rated "Inadequate."
Partially Inadequate	If at least one of B1 or B2 is rated as "Partially Inadequate," but neither is rated as "Inadequate."
Partially Adequate	At least one of B1 or B2 is rated "Partially Adequate," while neither is rated lower.
Adequate	If B1 AND B2 are rated as "Adequate."
Exceed	If B1 AND B2 are rated at least as "Adequate," and at least B1 or B2 are rated as "Exceed."

Use all evidence to rate components C1, C2, and C3.

Rating C1.

Rating Scale: C1 Overall	Criteria
Inadequate	At least one element of the strategy required to fulfill (B) above (i.e., necessary to address the threat) is likely to *not be legally feasible.*
Partially Adequate	There is some doubt whether all the elements of the strategy, which are required to fulfill (B) above (i.e., necessary to address the threat) *are legally feasible.*
Adequate	All the elements of the strategy that are required to fulfill (B) above (i.e., necessary to address the threat) *are also legally feasible.*

Rating C2.

Rating Scale: C2	Criteria
Inadequate	At least one element of the strategy required to fulfill (B) above (i.e., necessary to address the threat) is likely to *not be fiscally feasible.*
Partially Adequate	There is some doubt whether all the elements of the strategy, which are required to fulfill (B) above (i.e., necessary to address the threat) *are fiscally feasible.*
Adequate	All the elements of the strategy that are required to fulfill (B) above (i.e., necessary to address the threat) *are also fiscally feasible.*

Rating C3.

Rating Scale: C3	Criteria
Inadequate	At least one element of the strategy required to fulfill (B) above (i.e., necessary to address the threat) is likely to *not be politically feasible.*
Partially Adequate	There is some doubt whether all the elements of the strategy, which are required to fulfill (B) above (i.e., necessary to address the threat) *are politically feasible.*
Adequate	All the elements of the strategy that are required to fulfill (B) above (i.e., necessary to address the threat) *are also politically feasible.*

Combine components to Rate Overall Adequacy of C.

Rating Scale: C Overall	Criteria
Inadequate	At least one of C1, C2, or C3 is rated as "Inadequate."
Partially Adequate	At least one of C1, C2, or C3 is rated as "Partially Adequate," but none are rated as "Inadequate."
Adequate	C1, C2, and C3 are rated as "Adequate."

Combine A, B, and C to Yield Final Adequacy Rating.

Rating Scale: Final Adequacy	Criteria
Inadequate	Either A, B, or C are rated as "Inadequate."
Partially Inadequate	A or B is rated as "Partially Inadequate," and none of A, B, or C are rated as "Inadequate."
Partially Adequate	A, B, and C are all rated at least as "Partially Adequate," but not all are rated as "Adequate" or above.
Adequate	A, B, and C are all rated as "Adequate."
Exceed	A and C are rated as "Adequate," and B is rated as "Exceed."

1. After scoring the **Final Adequacy:** Answer the following questions, and provide additional support from external resources like literature, briefing presentations, your expertise, congressional hearings, etc.
   a. Identify
      i. What technical, policy, or resource gaps, if any, are limiting the strategy from being used to adequately identify international chemical threats? national chemical threats? Critical emerging threats?
   b. Prevent/Counter
      i. What technical, policy, or resource gaps, if any, are limiting the strategy from being used to adequately prevent nonstate actors from acquiring or misusing the technologies, materials, and critical expertise needed to carry out chemical attacks (including dual-use technologies, materials, and expertise? State-sponsored actors?
      ii. What technical, policy, or resource gaps, if any, are limiting the strategy from being used to adequately counter efforts by nonstate actors to carry out such chemical attacks? State-sponsored actors?
   c. Response
      i. What technical, policy, or resource gaps, if any, are limiting the strategy from being used to adequately respond to chemical terrorism incidents to attribute their origin and to help manage their consequence?

Appendix E

International Case Studies

AUM SHINRIKYO POISONING

The Aum Shinrikyo conducted 10 chemical attacks between 1990–1995. Summary descriptions of the three most important are considered here because they are relevant to the issue of identification. It is worthwhile noting that there remain aspects of Aum's activities that are not well understood, and so a comprehensive account does not exist; however, the summary provided by Danzig et al. (2012) provides additional detail. Aum was unusual in that it possessed significant scientific and chemical engineering expertise that included chemists with advanced degrees (Tucker, 2006). The cult was able to synthesize both VX and sarin, the latter in multiple-kilogram quantities, utilizing a custom-fabricated laboratory equipped with a modest level of computer control and air handling. The capability that Aum developed was remarkable, because normally the resources required to produce kilogram quantities of a nerve agent are significant, and would ordinarily require the backing of a nation-state, not a terrorist cult. The point is that terrorist organizations can be extremely well organized, and can have staff with significant technical training.

Aum Shinrikyo conducted two attacks using sarin. Neither attack was identified beforehand. The first occurred on June 27, 1994, in Matsumoto, Japan, in which sarin was released by volatilizing the liquid compound on a hot plate situated in the back of a truck and then blowing the resulting vapor out a window using a fan (Tucker, 2006). The cloud of vapor was blown by the wind to an apartment building, killing five residents. Matsumoto Emergency Services were notified by a resident who had been exposed. The apartment was evacuated, and 54 people were admitted to local hospitals. Blood tests indicated low levels of cholinesterase, and they were treated for organophosphate (OP) poisoning. There was no indication that the first responders recognized the release of the nerve agent. However, the physicians correctly identified OP poisoning based on the results of the blood tests.

As of June 28, local police investigators had not identified the poison. Finally, on July 3, a chemical analysis conducted by the Nagano Police Science Investigation Institute identified sarin breakdown products, six days after the attack. The Aum Shinrikyo perpetrators were not identified before the attack; their activities in procuring chemical processing equipment and organophosphorus compounds used to synthesize sarin had not attracted the attention of authorities.

Four months later, a leak in Aum's chemical processing apparatus resulted in contamination outside the building used for sarin synthesis. Chemical analysis of soil samples collected from near the building identified methyl phosphonic acid, which is a sarin degradation product. Other forensic research showed that Aum had procured significant quantities of chemicals that are precursors in the production of sarin. Despite the fact that these studies strongly implicated Aum in the attack on the apartment building, law enforcement (LE) elected not to confront Aum largely because the group was protected under the Japanese Religious Corporation Law, which prohibited investigation of registered religious groups "activities or doctrine" (Senate Government Affairs Permanent Subcommittee on Investigations, 1995). The rationale behind this decision is not known but was probably influenced by the lack of laws in Japan that prohibit the manufacture of chemical warfare agents, and perhaps an unwillingness to confront Aum, which displayed a combative and litigious response to any challenge to their activities. In retrospect, this represents a serious oversight in interdicting the threat before additional attacks occurred.

Additional chemical attacks did occur. In the fall of 1994, Aum members attempted to kill multiple individuals who opposed the cult using VX (James Martin Center for Nonproliferation Studies, 2016). In December of 1994, Aum cultists assassinated a former cult member in Osaka by applying drops of VX to his neck. He died in the hospital several days later, and the cause of death was not recognized by the Osaka police, nor, presumably, by the Osaka University Hospital staff.

The attack on the Tokyo subway occurred at about 8:00 a.m. on March 20, 1995. Plastic bags containing dilute sarin were punctured, which resulted in puddles on the floor of the subway, and subsequent evaporation of sarin. Hospitals in Tokyo were visited by 3,227 victims, and 493 of these were admitted (Tucker, 2006). Paramedics did not recognize nerve agent poisoning, and neither did emergency physicians at the hospitals.

Fifteen underground stations were affected. First responders—police, ambulance, and firefighters—entering the subway stations encountered commuters exiting the subway, and the first responders did not recognize cholinesterase poisoning in the chaos; consequently, many of the first responders were exposed to sarin. In addition to direct exposure, more than 200 ambulances and hospital staff received secondary exposure as a result of transfer from the clothing of the victims, another consequence of the inability of the emergency personnel to recognize the toxin. The cause of the incident was not immediately understood. By 11:00 a.m., the National Research Institute of Police Science identified sarin but did not inform the hospitals, who finally found out the identity of the poison via television. Dr. Nobuo Yanagisawa, who had treated victims of the

Matsumoto incident, recognized the symptoms while viewing media reports on television, and subsequently informed Tokyo hospitals (Farley, 1995). Identification of sarin by the hospitals was within hours after the first casualties arrived (Smithson, 2000).

Conclusions related to the identification stories topic:

1. First responders quickly concluded that a poison was responsible; however, they did not recognize cholinesterase poisoning and did not have training or equipment to prevent collateral exposures.
2. Hospital staff began to suspect sarin poisoning within hours of the arrival of the first casualties. However, there was no timely communication from police laboratories to the hospitals regarding the identification of sarin.
3. Timely communication to the public regarding the nature of the attack and the risk to the general public was lacking. "What brought the Tokyo hospital system under such pressure was not the truly injured, which hospitals proved more than capable of handling, but the monsoon of psychogenic patients" (Farley,1995). If correct information could have been more rapidly disseminated then pressures on responders and hospitals may have been significantly reduced.

SKRIPAL POISONING

Sergei and Yulia Skripal were poisoned in the United Kingdom by Russian agents using Novichok agent A234 (Carroll, 2018; Peplow, 2018; Haslam et. al., 2022), which is an acetyl cholinesterase inhibitor. The compound was applied to the outside door handle of their apartment (BBC News, 2018a). Exposure occurred when the Skripals touched the door handle (BBC, 2018; The Telegraph, 2018), which resulted in contamination of their skin. The compound functions by permeating into the skin, and diffusing into the vascular system, resulting in acetyl cholinesterase inhibition. The permeation and diffusion processes are relatively slow, and so a couple of hours elapsed before symptoms were manifest, during which time the Skripals had visited The Mill pub and Zizzi restaurant in Salisbury, finally moving to a park bench where they collapsed in response to the poisoning (BBC News, 2018b).

The couple was noticed by 16-year-old Abigail McCourt, who thought Sergei had suffered a heart attack. She alerted her mother Alison McCourt, who is an army colonel and chief nursing officer of Queen Alexandra's Royal Army Nursing Corps (*The Guardian*, 2019a). It is unclear whether Colonel McCourt recognized symptoms of nerve agent poisoning, and it is likely that the majority of first responder personnel were likely not trained to recognize nerve agent poisoning, particularly releases of next-generation agents like those used in the Skripals's poisonings.

This concern is substantiated by the exposure of the first responders, who reported itching eyes and breathing difficulties (*The Guardian*, 2019b). A total of 21 people (including the Skripals) were checked (itvNEWS, 2018). One police officer, Nick Bailey, received an exposure serious enough to warrant treatment in the intensive care unit. Two other police officers displayed minor symptoms. Officer Bailey was poisoned while he inspected the Skripals's house. This indicates that the forensic first responders were not apprised nor did they recognize the possibility of a nerve agent. It is not known whether Colonel McCourt informed first responders of the possibility of nerve agents.

Subsequently, British nationals Charlie Rowley and Dawn Sturgess were poisoned (BBC News, 2018c) as a result of contact with a sample of the same Novichok agent from a perfume bottle that had been used to transport the agent by the Russian agents. The perfume bottle had been discarded in a litterbin, found by Rowley, who gave it to Sturgess. She sprayed some of the contents of the bottle on her wrist, which resulted in her death eight days later. Rowley also was poisoned.

Paramedics responded; however, the police "initially thought the two patients had been using heroin or crack cocaine from a contaminated batch of drugs" (BBC News, 2018c). Police and firefighters who responded were in hazard suits and cordoned the area off, however, police did not declare a major incident until four days later, and shortly thereafter Scotland Yard concluded that Novichok was to blame following analysis at Porton Down (BBC News, 2018c). Reports indicate that the couple displayed symptoms consistent with nerve agent poisoning, (i.e., Sturgess was foaming at the mouth, and Rowley had pinpricked eyes, accompanied by sweating, and drooling). Yet it seems that these symptoms were not recognized by first responders, treatment personnel at the hospital, or LE until what was likely a few days later, despite widespread knowledge of the Skripal poisoning.

REFERENCES

BBC News. 2018a. "Spy Poisoning: Highest Amount of Nerve Agent Was on Door." March 18, 2018. https://www.bbc.com/news/uk-43577987.

BBC News. 2018b. "Russian Spy Poisoning: What We Know So Far." October 8, 2018. https://www.bbc.com/news/uk-43315636.

BBC News. 2018c. "Amesbury Novichok Poisoning: Couple Exposed to Nerve Agent." July 5, 2018. https://www.bbc.com/news/uk-44719639.

Carroll, Oliver. 2018. "Novichok Inventor on Amesbury Poisoning: 'I Completely Understand Panic of Those Living in Salisbury.'" *Independent*, July 6, 2018 https://www.independent.co.uk/news/uk/crime/novichok-inventor-amesbury-poisoning-salisbury-russia-vladimir-uglev-a8432876.html.

Danzig, R., M. Sageman, T. Leighton, L. Hough, H. Yuki, R. Kotani, and Z. Hosford. 2012. *Aum Shinrikyo: Insights into How Terrorists Develop Biological and Chemical Weapons*. 2nd ed. Center for a New American Security.

Farley, M. 1995. "Luck Played Role in Aid for Victims." *Los Angeles Times*, March 29, 1995. https://www.latimes.com/archives/la-xpm-1995-03-29-mn-48410-story.html

The Guardian. 2019a. "Novichok Poisoning Victims First Helped by Teenage Girl." January 20, 2019. https://www.theguardian.com/uk-news/2019/jan/20/novichok-poisoning-victims-sergei-skripal-first-helped-by-teenage-girl.

The Guardian. 2019b. "How Salisbury Case Went from Local Drama to International Incident." March 10, 2019. https://www.theguardian.com/uk-news/2018/mar/10/salisbury-poisoning-sergei-skripal-local-news-international-incident.

Haslam, J. D., Paul R., Stephanie H., Stevan R. E., and Peter G. B. 2022. "Chemical, Biological, Radiological, and Nuclear Mass Casualty Medicine: A Review of Lessons from the Salisbury and Amesbury Novichok Nerve Agent Incidents." *British Journal of Anaesthesia* 128, no. 2. https://doi.org/10.1016/j.bja.2021.10.008.

itvNEWS. 2018. "Up to 21 People Treated after Nerve Agent Attack on Russian Spy Sergei Skripal." March 8, 2018. https://www.itv.com/news/2018-03-08/nerve-agent-attack-state-russian-spy-daughter.

James Martin Center for Nonproliferation Studies. 2016. "Chronology of Aum Shinrikyo's CBW Activities." https://www.nonproliferation.org/wp-content/uploads/2016/06/aum_chrn.pdf

Peplow, M. 2018. "Nerve Agent Attack on Spy Used 'Novichok' Poison." *Chemical & Engineering News*, March 16, 2018.

Senate Government Affairs Permanent Subcommittee on Investigations. 1995. "Global Proliferation of Weapons of Mass Destruction: A Case Study on the Aum Shinrikyo."

Smithson, Amy E. 2000. Chapter 3 – "Rethinking the Lessons of Tokyo." 2000. *Ataxia: The Chemical and Biological Terrorism Threat and the US Response*. Washington, DC: Publication of the Henry L. Stimson Center

The Telegraph. 2018. "Russia Hacked Yulia Skripal's Emails For Five Years and Tested Novichok on Door Handles, Bombshell Intelligence Dossier Reveals." April 13, 2018. https://www.telegraph.co.uk/politics/2018/04/13/russia-hacked-yulia-skripals-emails-five-years-tested-novichok/

Tucker, J. B. 2006. *War of Nerves*. New York: Anchor Books.

Appendix F

Threats Interdicted Case Studies

DALLAS NATURAL GAS PLANT

The FBI, working as part of the Dallas Joint Terrorism Task Force (a coalition of federal and local police) arrested four individuals who were planning to blow up a natural gas processing plant in 1997. The group believed large quantities of hydrogen sulfide would be released (Pressley, 1997; Verhove, 1997). Identification relied on an informant who was part of the group, which did not have a formal name, but one member was associated with the Ku Klux Klan.

JAMES BELL

In 1997, a man from Oregon named James Bell was arrested for advocating attacks on federal agents, specifically IRS personnel.[1] When Bell was apprehended, he was found to have a significant quantity of sodium cyanide, although this seems to be a relatively minor aspect of what he was planning (AP, 1997). He was identified by an undercover IRS agent (Ryen, 1997) who infiltrated a far-right, sovereign citizen organization, the fake Multnomah County Common Law Court (which is not affiliated with the local or county courts) (Linzer and Rosenberg, 1997). Bell was arrested before he could carry out any attacks.

NEW ORDER

An Illinois-based white supremacist group called the New Order was arrested by the FBI in 1998, for planning to assassinate a lawyer from the Southern Poverty Law

---

[1] "Jim Bell," Wikipedia website, https://en.wikipedia.org/wiki/Jim_Bell.

Center (*New York Times*, 1998). The group also had aspirations to poison the water supplies of major cities.

CHARLES KILES AND KEVIN PATTERSON

In California, Charles Kiles and Kevin Patterson, from California—who were members of the San Joaquin militia—were arrested for plotting to use cyanide and explode propane tanks in 1999 (*Chicago Tribune*, 1999). Federal agents arrested two antigovernment militia members in connection with an alleged plan, after a nearly yearlong investigation by an FBI terrorism task force into a potential threat against the Suburban Propane facility in Elk Grove and other targets in the Sacramento area.

DEMETRIUS VAN CROCKER

In 2004, Demetrius Van Crocker, from Tennessee, wanted to acquire sarin and said he made mustard agent (though there is no evidence he did) as part of a plot targeting government facilities (FBI, 2006). Crocker attempted to acquire the nerve agent and C4 explosives from an undercover FBI agent. A concerned citizen took his "extremist rants" and plans "to build a dirty bomb to blow up a state or federal courthouse" (FBI, 2006) seriously enough to call the Tennessee Bureau of Investigation, which in turn called the FBI. "We thought there might be something to it," said Special Agent Daryl Berry (FBI, 2006), who opened the case in September 2004 out of the FBI's office in Jackson, Tennessee. That set in motion the undercover sting that resulted in the apprehension of Van Crocker.

MYRON TERESHCHUK

Also in 2004, the FBI apprehended Myron Tereshchuk, who was being investigated as part of a separate extortion case using the internet (*Washington Post*, 2004). When law enforcement searched his home as part of the investigation, materials for making hand grenades and "items necessary for making [extracting] ricin," along with "literature about poisons" were seized (DOJ, 2004). Ricin is a naturally occurring highly toxic protein (i.e., a biomacromolecule), and hence lies at the convergence of biology and chemistry, which complicates the understanding of how to categorize the threat; however, identification of the threat would involve the same government organizations. It is not clear that it can be inferred that he had the knowledge needed to know how to extract ricin. Given the specifics noted in the Department of Justice (DOJ) indictment, it is reasonable to infer that he may have had castor beans, which are readily accessible, and some equipment. Because he was not charged with possession of ricin, LE may not have been able to detect any significant amounts of the protein. Again, motives are not readily available. This case may be illustrative of serendipity, which is problematic as a strategy: there is no publicly available information indicating Tereshchuk was being investigated on suspicion of terrorism in general or chemical terrorism specifically.

JEFFREY DETRIXHE

In 2008, Jeffrey Detrixhe attempted to sell sodium cyanide to an FBI informant (DOJ, 2008; ABC News, 2008) in connection with a case being investigated in the context of illegal activities by a right-wing, white supremacist group. Detrixhe had acquired a 25-gallon drum containing 62 pounds of cyanide, which he allegedly offered to sell for $10,000, a thermal imager, and an assault rifle (DOJ, 2008). It is not clear that Detrixhe intended to do anything with the sodium cyanide himself; he was convicted of possession of a substance without a legitimate purpose. The intended recipient, "fat Bob," was a member of the Aryan Brotherhood, and Detrixhe acknowledged that "he had second thoughts about selling it [the sodium cyanide] because it would probably be used for a bad purpose" (DOJ, 2008). How Detrixhe acquired a large amount of sodium cyanide remains unanswered.

RYAN CHAMBERLAIN

Another example of a potential threat being interdicted is the 2014 case of Ryan Chamberlain, from California, who had procured rosary peas and sodium cyanide (Dinzeo, 2016). FBI agents were made aware that Chamberlain was accused of working toward isolating a biological toxin, abrin. Abrin can be isolated from the bright red and black seeds, which are known as rosary peas or jequirity beans, of the invasive pantropical *Abrus precatorius* plant. He also had attempted to purchase abrin, ricin, and nicotine via an online black market seller (Business Insider, 2014). In addition to crushed rosary peas, Chamberlain was also found to possess explosive-making materials and had removed the serial number from a firearm. Reportedly, Chamberlain was apprehended before he developed a specific plot for an attack, where his desire was to create a toxic powder that could be widely dispersed to harm others. The motive behind Chamberlain's intention to use abrin remains unclear. There have been some suggestions he thought the end times as suggested in the Biblical Book of Revelations was about to occur. The allegations regarding the scale of pursuit of chemical terrorism were further muddied when the U.S. district judge in the case noted it was "apparent that the U.S. Attorney's Office never had any reliable basis for asserting that the FBI recovered between 1,000 and 2,000 lethal doses of abrin from Chamberlain's apartment" (Dinzeo, 2016). The lack of clarity surrounding motive, what was actually pursued, and the extent of actual agent obtained makes this a difficult case from which to draw conclusions with any confidence.

JARRETT WILLIAM SMITH

The case against Jarrett William Smith, who pled guilty to two counts of "distributing information related to explosives, destructive devices and weapons of mass destruction" (DOJ, 2020), to possible right-wing extremist groups, is illustrative of the challenges of identifying domestic violent extremists with interest in chemical terrorism. While serving on active duty in the U.S. Army, Smith, between 2017–2019, became self-radicalized (DSCA, 2019). The group Smith joined linked him to the Order of Nine Angles (O9A), which is a leaderless, decentralized network that espouses white supremacist, neo-Nazi, pro-jihadist ideas with a shared neo-Satanist ideology, and actively promotes violent terrorism (Koch, 2022). This group has been in the news recently in conjunction with another former soldier sentenced to 45 years in prison for "attempting to murder U.S. service members, providing and attempting to provide material support to terrorists, and illegally transmitting national defense information," passing operational deployment information to other members of the O9A (DOJ, 2023). Jarrett Smith was charged with Distributing Information Relating to Explosives, Destructive Devices, and Weapons of Mass Destruction, in violation of Title 18, United States Code, Section 842(p) (CourtListener, 2019a).

To interdict the chemical threat posed by Smith, the FBI initially received information on comments Smith had posted on social media about his intent "to travel to Ukraine to fight with a violent, far-right military group" (DOJ, 2020). How the FBI was made aware of those comments has not been made public; in previous instances, representatives of Meta, the parent company of Facebook, have indicated they shared information with federal LE when intent to commit politically motivated violence was suspected (GAO, 2022). Additionally, a confidential informant provided further information that eventually led the FBI to act. This information included Smith's offer of what he claimed were methods for producing chemical agents intended to injure or kill domestic elected officials. The charging complaint, specifically, provides a transcript of Smith's statements that included instructions to generate chlorine and what was subsequently identified as napalm[2]:

> *Smith: If you want a quick and cheap gas grenade, a [combination of commonly available chemicals] will work. [Instructions for activating the device]. Blows in 8–15 seconds. One hell of a wallop and it leaves behind a cloud of toxic chlorine gas* (CourtListener, 2019a).

> *Smith: Ok. I think I have an idea for you. You will need [various household chemicals and commonly available equipment]. You can keep all the materials separate until it's time. Plus the randomness will aid you in the case of searches and the materials themselves usually aren't considered suspicious* (CourtListener, 2019b).

The FBI's criminal complaint contains two additional instances of Smith suggesting the use of chemicals to commit violence. Smith repeatedly emphasized how accessible the starting materials were:

> FBI Undercover Employee: I am reading and thinking but this looks really good. I like the fact that everything is stuff you find around the house.

---

[2] NB: Napalm is not a chemical weapon under the Chemical Weapons Convention (CWC). Under international law it is considered an incendiary when used against military targets; under the UN Convention on Certain Conventional Weapons (CCW), use against civilian populations is prohibited. Similarly, under the Convention on the Prohibition of Military or Any Other Hostile Use of Environmental Modification Techniques (ENMOD treaty), use of napalm as an incendiary defoliant is also prohibited.

Smith: That's the best way to fight people. Making AK-47s out of expensive parts is cool, but imagine if you will if you were going to Walmart instead of a gun store to buy weapons (CourtListener, 2019b).

When arrested, Smith stated that his goal was "to cause 'chaos,'" (CourtListener, 2019c) which is consonant with "accelerationism"–the acceleration of sociopolitical collapse via acts of violence and terrorism–goals of the O9A with which he claimed affiliation (Koch, 2022).

REFERENCES

ABC News. 2008. "Ryan, Jason, Texas Man Accused of Trying to Sell Cyanide." May 14, 2008. https://abcnews.go.com/TheLaw/FedCrimes/story?id=4856904&page=1.

AP (The Associated Press). 1997. "Bell Gets 11 Months in Prison, 3 Years Supervised Release, Fine." December 12, 1997. https://cryptome.org/jdb/jimbell7.htm.

Business Insider. 2014. "California Consultant Was Arrested for Trying to Buy Toxins Online." https://www.businessinsider.com/r-arrested-california-consultant-sought-toxins-online-fbi-2014-07.

Chicago Tribune. 1999. "FBI: Militia Plotted to Bomb Propane Tanks." *Chicago Tribune.* Retrieved from https://www.chicagotribune.com/news/ct-xpm-1999-12-05-9912050342-story.html.

CourtListener. 2019a. "Indictment Filed against Defendant Jarrett William Smith as to Counts 1 – 3," U.S. District Court, D. Kansas. September 25, 2019. p.1, https://www.courtlistener.com/docket/16246401/united-states-v-smith.

CourtListener. 2019b. United States v. Smith (5:19-cr-40091), District Court, D. Kansas. September 23, 2019, p.5. https://www.courtlistener.com/docket/16246401/united-states-v-smith.

CourtListener. 2019c. United States v. Smith (5:19-cr-40091), District Court, D. Kansas. September 23, 2019, p.7. https://www.courtlistener.com/docket/16246401/united-states-v-smith.

Dinzeo, M. 2016. "San Fran PR Man Pleads Guilty to Poison Charge." *Courthouse News Service,* February 16, 2016. https://www.courthousenews.com/san-fran-pr-man-pleads-guilty-to-poison-charge.

DOJ (U.S. Department of Justice). 2004. "Wi-Fi Hacker Pleads Guilty to Attempted $17,000,000 Extortion." United States Department of Justice website, June 8, 2004. https://www.justice.gov/archive/criminal/cybercrime/press-releases/2004/tereshchukPlea.htm.

DOJ. 2008. "Man Sentenced to Nearly Six Years in Federal Prison for Possessing Cyanide." U.S. Department of Justice website, United States Attorney Richard B. Roper, Northern District of Texas, December 17, 2008. https://www.justice.gov/archive/usao/txn/PreOssRel08/detrixhe_sen_pr.html.

DOJ. 2020. "Former Fort Riley Soldier Sentenced for Distributing Info on Napalm, IEDs." August 19, 2020. https://www.justice.gov/usao-ks/pr/former-fort-riley-soldier-sentenced-distributing-info-napalm-ieds.

DOJ. 2023. "Former U.S. Army Soldier Sentenced to 45 Years in Prison for Attempting to Murder Fellow Service Members in Deadly Ambush." March 3, 2023. https://www.justice.gov/usao-sdny/pr/former-us-army-soldier-sentenced-45-years-prison-attempting-murder-fellow-service.

DSCA (Defense Counterintelligence and Security Agency). 2019. "Center for Development of Security Excellence (CDSE) Case Study: Self-Radicalization." https://www.cdse.edu/Portals/124/Documents/casestudies/case-study-smith.pdf.

FBI (Federal Bureau of Investigation). 2006. "Domestic Terrorism Tips Lead to Sting, Prison for Plotter." The FBI Federal Bureau of Investigation website, November 2006. https://archives. fbi.gov/archives/news/stories/2006/november/terror_112906.

GAO. 2022. "Capitol Attack: Federal Agencies' Use of Open Source Data and Related Threat Products Prior to January 6, 2021," GAO-22-105963, May 02, 2022, p. 20. https://www. gao.gov/products/gao-22-105963.

Koch, A. 2022. "The ONA Network and the Transnationalization of Neo-Nazi-Satanism." *Studies in Conflict & Terrorism.* https://doi.org/10.1080/1057610X.2021.2024944.

Linzer, L., and D. Rosenberg. 1997. *Vigilante Justice: Militias and "Common Law Courts" Wage War Against the Government.* Anti-Defamation League. https://adl.org/sites/default/files/ documents/assets/pdf/combating-hate/adl-report-1997-vigilante-justice.pdf.

New York Times. 1998. "Supremacists Had Hit List, F.B.I. Agent Says." *The New York Times* website, March 7, 1998. https://www.nytimes.com/1998/03/07/us/supremacists-had-hit-list-fbi-agent-says.html.

Pressley, S. A. 1997. "Group Planned Massacre and Big Robbery, FBI Says." *The Washington Post* webpage, April 25, 1997. https://www.washingtonpost.com/archive/politics/1997/04/25/ group-planned-massacre-and-big-robbery-fbi-says/8d864957-b2cf-4c39-be2a-10361b6c31de.

Ryen, J. V. 1997. *Digital Files from Court Reporter Julaine V. Ryen.* Tacoma, WA: Western District of Washington. https://cryptome.org/usa-v-jdb-01.htm.

Verhovek, S. H. 1997. "U.S. Officials Link Klan Faction to 1 of 4 People Held in Texas Bomb Plot." *The New York Times* website, April 25, 1997. https://www.nytimes.com/ 1997/04/25/us/us-officials-link-klan-faction-to-1-of-4-people-held-in-texas-bomb-plot.html.

Washington Post. 2004. "Ricin and Grenades Found at MD Home." September 29, 2004 https:// www.washingtonpost.com/archive/local/2004/09/30/ricin-and-grenades-found-at-md-home-in-extortion-probe/c08045be-fe71-4439-a789-d9d45f3c988f.

Appendix G

Threats Manifested Case Studies

This review of domestic threats of terrorism using chemical materials highlights some cases in which the intelligence community (IC) and law enforcement (LE) were able to interdict terrorism threats before they could be conducted. However, in the majority of the incidents reviewed here, a detailed account of how the threat was initially identified was not publicly available. When the mode of threat identification could be discerned, a combination of informants, perpetrator incompetence, state and local LE, and multiple federal agencies (notably the Secret Service, Customs) acting together with the FBI were involved. In some instances, threat identification points to a high level of expertise, coordination, and efficiency; in other cases, some degree of luck and fortuitous happenstance were involved. These considerations illustrate the reality that the "identify" task is necessarily not systematic, which indicates the need for an agile, robust communication structure. Furthermore, not all threats have been interdicted, an observation substantiated by the cases described here and some State-based targeted assassination attempts such as the Skripal poisoning (see Appendix E).

JOSEPH LORIS

A threat not identified prior to an attack being carried out was perpetrated by Joseph Loris in 2014, who attacked the Social Security Building in Santa Cruz, CA (Santa Cruz Police Department Blog, 2014). He poured ammonia and Clorox bleach into the building, which mixed to create chloramine gas. It is not clear that Loris meant to produce a gas. Loris was homeless and possibly suffering from mental illness. After the attack, Loris was identified and apprehended by Santa Cruz Police Detectives, with assistance from the FBI. The episode illustrates the impossibility of identifying all threats before they are carried out and some of the risks inherent in household materials.

WORLD TRADE CENTER BOMBING

The 1993 bombing of the World Trade Center (WTC) in New York City was an event in which the threat was not identified before the attack was carried out, but the event does not qualify as a chemical attack either. A group of five members of al-Qa'ida, led by Ramzi Yousef, drove a van containing a bomb into a parking garage under the WTC and detonated it (Parachini, 2000). Six people were killed, and more than 1,000 were injured. The attack was not a chemical attack; however, Yousef "seriously considered employing chemical agents" in the WTC bombing. The perpetrators had procured a small quantity of sodium cyanide. The FBI apprehended all of the perpetrators except Yousef, who was eventually caught by Philippine police nearly two years after the attack.

REFERENCES

Parachini, J. V. 2000. "The World Trade Center Bombers (1993)." In *Toxic Terror: Assessing Terrorist Use of Chemical and Biological Weapons*, edited by Jonathan B. Tucker, 185–206. Cambridge, MA: MIT Press.

Santa Cruz Police Department Blog. 2014. "Arrest Made in Chemical Attack on Social Security Office." May 20, 2014. http://santacruzpolice.blogspot.com/2014/05/arrest-made-in-chemical-attack-on.html.